Ernst Schering Research Foundation Workshop
Supplement 4
Hormone Replacement Therapy and Osteoporosis

Springer-Verlag Berlin Heidelberg GmbH

Ernst Schering Research Foundation Workshop
Supplement 4

Hormone Replacement Therapy and Osteoporosis

J. Kato, H. Minaguchi, Y. Nishino
Editors

With 74 Figures and 15 Tables

Springer

Series Editors: G. Stock and M. Lessl

ISSN 0947-6075
ISBN 978-3-662-04023-2

CIP data applied for

Die Deutsche Bibliothek – CIP-Einheitsaufnahme
Schering-Forschungsgesellschaft <Berlin>:
[Ernst Schering Research Foundation Workshop / Supplement].
Ernst Schering Research Foundation Workshop. Supplement. - 1-...

Erscheint unregelmäßig. - Bibliographische Deskription nach 5 (1999)
Reihe Supplemnt zu: Schering-Forschungsgesellschaft <Berlin>: Ernst Schering Research Foundation Workshop
4. Hormone replacement therapy and osteoporosis. - 2000
Hormone replacement therapy and osteoporosis / Ernst Schering Research Foundation. J. Kato ... ed.

(Ernst Schering Research Foundation Workshop : Supplement; 4)
ISBN 978-3-662-04023-2 ISBN 978-3-662-04021-8 (eBook)
DOI 10.1007/978-3-662-04021-8

Originally published by Springer-Verlag Berlin Heidelberg New York in 2000
Softcover reprint of the hardcover 1st edition 2000

Typesetting: Data conversion by Springer-Verlag

SPIN:10691243 21/3134/AG–5 4 3 2 1 0 – Printed on acid-free paper

Preface

This book presents the Proceedings of the International Joint Workshop of Nihon Schering and the Ernst Schering Research Foundation on "Hormone Replacement Therapy," which took place on 9–11 November 1998 in Osaka, Japan.

In cases of advanced age, hormone replacement therapy (HRT) is becoming increasingly important both for women's health and for the doctors in our profession. In-depth understanding of HRT is needed. The organizers and editors intended the workshop to consist of six sessions, as follows:

1. Overview of HRT
2. Basic aspects of steroid hormones
3. HRT and the brain
4. The influence of HRT on lipids and the cardiovascular system
5. HRT and bone
6. HRT – its effects and risks

It was also intended that each session should be a combination of basic and clinical lectures.

Without a basic understanding of sex steroid hormones in targets such as the reproductive organs, including the breast and uterus, and the brain, as well as non-targets (bone), possibilities for improving treatment would not emerge. This we firmly believe.

The front-runners from at home and abroad in this field joined the workshop to provide the most advanced knowledge and state-of-the-art information, basic and clinical, regarding HRT.

The participants in the workshop

Active and useful discussions and comments also emerged during the course of the workshop.

Junzo Kato, Hiroshi Minaguchi, Yukishige Nishino

Table of Contents

List of Editors and Contributors

Editors

J. Kato
Teikyo Heisei Junior College, 6–29–1 Ichiharadai, Ichihara-shi,
Chiba, 290–0192, Japan

H. Minaguchi
Minaguchi Hospital, 1–7-7 Minamicho, Kichijoji Musashino-shi,
Tokyo, 180–0003, Japan

Y. Nishino
Preclinical Development Department, Nihon Schering K.K.,
6–64, Nishimiyahara 2-Chome, Yodogawa-ku, Osaka-shi,
Osaka, 532–0004, Japan

Contributors

K. Akazawa
Dokkyo Medical University, Department of Obstetrics and Gynecology,
Koshigaya Hospital, 2–1-50 Minami-koshigaya, Koshigaya-shi,
Saitama, 343–8555, Japan

T. Aono
Tokushima University, School of Medicine,
Department of Obstetrics and Gynecology, 3–18–15 Kuramoto-cho,
Tokushima-shi, Tokushima, 770–8503, Japan

T. Aso
Tokyo Medical and Dental University, School of Medicine,
Department of Obstetrics and Gynecology, 1–5-45 Yushima, Bunkyo-ku,
Tokyo, 113–0034, Japan

K. Azuma
Tokushima University, School of Medicine,
Department of Obstetrics and Gynecology, 3–18–15, Kuramoto-cho,
Tokushima-shi, Tokushima, 770–8503, Japan

M. Hamamoto
Dokkyo Medical University, Department of Obstetrics and Gynecology,
Koshigaya Hospital, 2–1-50 Minami-koshigaya, Koshigaya-shi,
Saitama, 343–8555, Japan

H. Hiroi
Saitama Medical School, Department of Biochemistry, 38 Morohongo,
Moroyama-machi, Irima-gun, Saitama, 350–0495, Japan

H. Honjo
Kyoto Prefectural University of Medicine,
Department of Obstetrics and Gynecolgy, 465 Kajii-cho, Hirokojiagaru,
Kawaramachidori, Kamigyo-ku, Kyoto, 602–0841, Japan

K. Ikeda
Saitama Medical School, Department of Biochemistry, 38 Morohongo,
Moroyama-machi; Irima-gun, Saitama, 350–0495, Japan

S. Inoue
Saitama Medical School, Department of Biochemistry, 38 Morohongo,
Moroyama-machi, Irima-gun, Saitama, 350–0495, Japan

M. Irahara
Tokushima University, School of Medicine,
Department of Obstetrics and Gynecology, 3–18–15, Kuramoto-cho,
Tokushima-shi, Tokushima, 770–8503, Japan

K. Isse
Dokkyo Medical University, Department of Obstetrics and Gynecology,
Koshigaya Hospital, 2–1-50 Minami-koshigaya, Koshigaya-shi,
Saitama, 343–8555, Japan

K. Iwasa
Kyoto Prefectural University of Medicine,
Department of Obstetrics and Gynecolgy, 465 Kajii-cho, Hirokojiagaru,
Kawaramachidori, Kamigyo-ku, Kyoto, 602–0841, Japan

N. Iwasaki
Dokkyo Medical University, Department of Obstetrics and Gynecology,
Koshigaya Hospital, 2–1-50 Minami-koshigaya, Koshigaya-shi,
Saitama, 343–8555, Japan

T. Kashiwagi
Kyoto Prefectural University of Medicine,
Department of Obstetrics and Gynecolgy, 465 Kajii-cho, Hirokojiagaru,
Kawaramachidori, Kamigyo-ku, Kyoto, 602–0841, Japan

A. Kikuchi
Department of Obstetrics and Gynecology, Faculty of Medicine,
University of Tokyo, Tokyo, 7–3-1, Hongo, Bunkyo-ku,
Tokyo 113–8655, Japan

N. Kikuchi
Kyoto Prefectural University of Medicine,
Department of Obstetrics and Gynecolgy, 465 Kajii-cho, Hirokojiagaru,
Kawaramachidori, Kamigyo-ku, Kyoto, 602–0841, Japan

H. Matsumi
Department of Obstetrics and Gynecology, Faculty of Medicine,
University of Tokyo, Tokyo, 7–3-1, Hongo, Bunkyo-ku,
Tokyo 113–8655, Japan

H. Minaguchi
Ryokuseikai Minaguchi Hospital, 1–7-7 Minamicho, Kichiyouji Musashino-shi, Tokyo, 180–0003, Japan

C. Miyaura
Tokyo University of Pharmacy and Life Science, Department of Biochemistry
School of Pharmacy, 1432–1 Horinouchi, Hachiouji, Tokyo, 192–0392, Japan

M. Muramatsu
Saitama Medical School, Department of Biochemistry, 38 Morohongo,
Moroyama-machi; Irima-gun, Saitama, 350–0495, Japan

H. Nakayama
Department of Obstetrics and Gynecology, Faculty of Medicine, University of Tokyo, Tokyo, 7–3-1, Hongo, Bunkyo-ku, Tokyo, 113–8655, Japan

M. Nozaki
Kyusyu University, School of Medicine, Department of Obstetrics and Gynecology, 3–1-1 Made, Higashi-ku, Fukuoka-shi, Fukuoka, 812–8582, Japan

S. Ogawa
Saitama Medical School, Department of Biochemistry, 38 Morohongo, Moroyama-machi, Irima-gun, Saitama, 350–0495, Japan

T. Ohkura
Dokkyo Medical University, Department of Obstetrics and Gynecology, Koshigaya Hospital, 2–1-50 Minami-koshigaya, Koshigaya-shi, Saitama, 343–8555, Japan

T. Okubo
Kyoto Prefectural University of Medicine, Department of Obstetrics and Gynecolgy, 465 Kajii-cho, Hirokojiagaru, Kawaramachidori, Kamigyo-ku, Kyoto, 602–0841, Japan

A. Orimo
Saitama Medical School, Department of Biochemistry, 38 Morohongo, Moroyama-machi, Irima-gun, Saitama, 350–0495, Japan

D.W. Pfaff
The Rockefeller University School of Medicine, 1230 York Avenue, New York, NY 10021, USA

H.P.G. Schneider
University of Münster, Department of Obstetrics and Gynecolgy, Albert-Schweitzer-Str. 33, 48149 Münster, Germany

M. Taga
Department of Obstetrics and Gynecology, Yokohama City University, School of Medicine, 3–9 Fukura, Kanazawa-ku, Yokohama, 236, Japan

E. Takatsu
Research Institute of Shiseido Cosmetics, 1050 Niibacho, Kohoku-ku, Yokohama-shi, Kanagawa, 223–0057, Japan

Y. Taketani
Department of Obstetrics and Gynecology, Faculty of Medicine, University of Tokyo, Tokyo, 7–3-1, Hongo, Bunkyo-ku, Tokyo, 113–8655, Japan

K. Tanaka
Dokkyo Medical University, Department of Obstetrics and Gynecology, Koshigaya Hospital, 2–1-50 Minami-koshigaya, Koshigaya-shi, Saitama, 343–8555, Japan

H. Tsuchiya
Kyoto Prefectural University of Medicine, Department of Obstetrics and Gynecolgy, 465 Kajii-cho, Hirokojiagaru, Kawaramachidori, Kamigyo-ku, Kyoto, 602–0841, Japan

M. Urabe
Kyoto Prefectural University of Medicine, Department of Obstetrics and Gynecolgy, 465 Kajii-cho, Hirokojiagaru, Kawaramachidori, Kamigyo-ku, Kyoto, 602–0841, Japan

Y. Wang
Department of Obstetrics and Gynecology, Faculty of Medicine, University of Tokyo, Tokyo, 7–3-1, Hongo, Bunkyo-ku, Tokyo 113–8655, Japan

T. Watanabe
Saitama Medical School, Department of Biochemistry, 38 Morohongo, Moroyama-machi, Irima-gun, Saitama, 350–0495, Japan

T. Yamamoto
Kyoto Prefectural University of Medicine, Department of Obstetrics and Gynecolgy, 465 Kajii-cho, Hirokojiagaru, Kawaramachidori, Kamigyo-ku, Kyoto, 602–0841, Japan

T. Yano
Department of Obstetrics and Gynecology, Faculty of Medicine,
University of Tokyo, Tokyo, 7–3-1, Hongo, Bunkyo-ku,
Tokyo, 113–8655, Japan

M.G. Zhang
Department of Obstetrics and Gynecology, Yokohama City University,
School of Medicine, 3–9 Fukuura, Kanazawa-ku, Yokohama, 236, Japan

1 General Aspects of Worldwide HRT Use

H.P.G. Schneider

1.1 Introduction

The general impression is that the human lifespan is increasing. The lifespan is the biological limit to life; it is the maximum age obtainable by a member of a species. There is a statistical variance from this for individuals, but overall, lifespan is finite. What is really increasing is the life expectancy rather than the lifespan. An analysis of the average life expectancy indicates that death converges at a certain maximum age. Thus, the number of older people is increasing, but it will eventually reach a fixed limit. The time spent in old age will make up an increasing percentage of a typical life, as we have nearly eliminated premature death. The reason for this success is a virtual eradication of previously fatal infectious diseases.

It is 1998, the second-last year of the twentieth century. This century has been filled with upheavals and advances. Some of the most exciting and beneficial advances have been those in human health. The last half of this century has indeed offered more advances in living conditions than have occurred in all the centuries before. Women and men of different racial and ethnic origins, cultures, ages, and socioeconomic backgrounds are part of this evolutionary process. Key issues of women's health include research and clinical practice, the enhanced role of women as informed consumers and patients, gender-specific education of scientists and general health providers, and the general acceptance of women's health as an integral part of our health system infrastructure.

1.2 Aging of Women

According to the World Health Report 1997, health expectancy can be defined as life expectancy in good health and amounts to the average number of years an individual can expect to live in such a favorable state. It is vital to realize that increased longevity does not come without costs. Every year, many millions die prematurely or are disabled by diseases and conditions that are to a large extent preventable. Longer life can be a penalty as well as a prize. A large part of the price to be paid is in the currency of chronic disease. Chronic diseases are responsible for more than 24 million deaths a year, or almost half of the global total. The leading causes are circulatory diseases, including heart disease and strokes, cancer, and chronic obstructive pulmonary disease. Chronic diseases, with a few exceptions, have not so far lent themselves so easily to cure. They are less open to community action. They do not spread from person to person. Every case of chronic disease represents a burden for one individual who, depending on circumstances, may or may not have access to treatment or support.

The world population reached a total of 5,800 million in mid-1996. It increased by more than 80 million during that year. The child and adolescent population grew by about 0.7%, the adult population by 1.8%, and that of the elderly by 2.4%. The number of people aged 65 years and above increased to 380 million, reflecting a 14% global increase in that age group between 1990 and 1995. It is projected that the global over-65 population will have increased by 82% between 1996 and 2020; this corresponds to approximately 110% in the least developed and developing countries, and approximately 40% in developed countries. In 1996, life expectancy at birth reached 65 years. It has increased globally by about 4.6 years between 1980 and 1995 (4.4 years for males and 4.9 years for females).

Of the more than 52 million worldwide deaths in 1996, over 17 million were ascribable to infectious and parasitic diseases; more than 15 million were due to circulatory diseases; over 6 million were caused by cancers; and about 3 million resulted from non-specific respiratory diseases. More than 10 million people developed cancer in 1996 and over 6 million of those who already had the disease died of it. As most cancers appear in adults who are at an advanced age, the burden of cancer is much more important than that of other diseases in populations with long life expectancy. The worldwide leading cancer killers, which together account for about 60% of all cancer cases and deaths, are cancers of the lung, stomach, breast, colon, rectum, mouth, liver, cervix, and esophagus. Although they do not share the same risk factors, a few major factors dominate this group, namely diet, tobacco, infections, alcohol, and, reportedly, hormones.

1.3 The Role of Preventive Health Care

Most of the remaining premature deaths are now concentrated in the years over the age of 60, and are due to chronic disease. Major clinical disciplines, such as internal medicine, surgery, and gynecology, concentrate more and more on the later years of life. An appropriate goal is to have healthy and independent elders who maintain physical and cognitive function as long as possible. Fries (1988) described recent eras of human development in terms of health and disease. The first era existed until some time in the early 1900s, and was characterized by identification and, ultimately, the successful therapy of infectious diseases. The second era, highlighted by cardiovascular disease and cancer, is now beginning to fade into the third era, which is marked by problems of frailty, diminution of function (fading eyesight and hearing, impaired cognitive function and memory), and decreased strength and stamina. The medical practice of "find the disease and cure it" still acts as if we were still in the initial era. The best health strategy towards improving chronic illness, however, is to change the slope, the rate at which illness develops, thus postponing the clinical illness, and if it is postponed long enough, effectively preventing it. The methods for postponing illness are obvious to us all: exercise, elimination of cigarette smoking, avoiding excessive alcohol consumption, elimination of obesity, and – particularly in the elderly – a sense of personal choice in dealing with individual problems. The individual sense of controlling health is significantly affected by aging (Rodin 1986). Studies involving older people emphasize the importance of self-determination. Older people with a sense of control are more likely to improve their cognitive function, seek answers, and adhere to programs of preventive health care.

Postponing illness is expressed by Fries et al. as *the compression of morbidity* (Fries 1980; Fries et al. 1989). We would live relatively healthy lives and compress our illnesses into a short period of time just before death around age 85. Disease is therefore something not necessarily best treated by medication or surgery, but by prevention, or, more accurately, by postponement.

The primary purpose of health promotion is to improve quality of life. Preventive health programs have a greater impact on morbidity than upon mortality (Fries et al. 1989). Currently, our greatest health problems remain with the senior population, and socioeconomic strategies should focus on health promotion in the elderly. The optimal choice is what Fries calls *linear senescence* or *maximizing the vigor in life*. Some linear decline is unavoidable, but the slope can be changed by effort and practice.

1.4 Worldwide HRT Use

Preventive health care education is important throughout life. At the time of menopause, however, a review of the major health issues can be extremely rewarding. This physiological event is a reason for many women to contact their physicians, thus providing the opportunity to enroll in a preventive health care program. The primary deficit at menopause is ovarian. The decline of estrogen is associated with the clinical signs and symptoms that are usually associated with menopause, as well as at least some acceleration in the development of coronary atherosclerosis and osteoporosis. In addition, the changing amount of androgen and the changes in estrogen–androgen ratios also cause clinically important alterations. Replacing these hormonal deficits has been recommended widely; this is in contrast to clinical strategies in other organ deficits such as in thyroid, adrenal, or pancreatic disease.

1.4.1 Clinical Concerns

Hormone replacement therapy (HRT) can largely prevent or mitigate the complications associated with menopause. Considerable attention has been paid to the relief of vasomotor symptoms, prevention of osteoporosis and cardiovascular morbidity and mortality, and the elimination of urogenital atrophy. The effects of HRT on mood, cognitive function, and risk of Alzheimer's disease and colo-rectal cancer are also promising.

Successful HRT requires an individualized regimen that meets the woman's specific needs and symptoms, by providing adequate estrogen levels with the lowest effective dose and with a route of administration that best suits her. The physician and patient need to confer to select which of the four basic HRT regimens is most appropriate:

1. Continuous estrogen alone (patients without uterus)
2. Cyclic estrogen plus sequential progestogen
3. Continuous estrogen plus sequential progestogen (patients have scheduled monthly bleeds)
4. Continuous combined estrogen and progestogen (minority of patients have unscheduled bleeds for six to eight months)

Numerous studies observed bleeding patterns that were obtained with various HRT regimens. Continuous combined HRT results in fewer bleeding episodes and has been associated with better long-term compliance than has been found with the sequential regimen (Dören et al. 1995; Eiken and Kolthoff 1995). At the end of an 8-year study, none of the women in the continuous HRT group changed to other therapies and 46% were still receiving treatment whereas only 32% remained on sequential HRT (Eiken and Kolthoff 1995).

Studies have shown that prolonged unopposed estrogen replacement therapy increases the risk of endometrial cancer. Several European and US studies have provided evidence, however, that adding a progestogen significantly reduces the risk of endometrial hyperplasia and carcinoma in postmenopausal women (Persson et al. 1989; Beresford et al. 1997). The addition of 12 and more days of a progestogen within a 28-day cycle in short- and longer-term HRT will reduce endometrial cancer risk more or less to baseline (Beresford et al. 1997); endometrial hyperplasia is seen in less than 1% of the cases.

Estrogens (conjugated equine estrogens and estradiol-based) in most of the cases give relief from flushes and night sweats, especially if these are frequent and severe. Progestogens have also been shown to be effective on symptoms when used in adequate daily dosages (e.g., MPA 10 mg; dydrogesterone 10 mg). Tibolone, 2.5 mg/day, given continuously reduces vasomotor symptoms. Tamoxifen is not effective at all and can even cause flushes; this is also true for second-generation SERMs such as raloxifene, which will dose-dependently increase menopausal flushes (Delmas et al. 1997).

Dysuria, urgency, urge incontinence, urinary frequency, and nocturia all increase with age. Almost 50% of the postmenopausal women with urinary incontinence relate the onset of their problems to menopause. There is a strong correlation between incontinence and other urogenital symptoms in postmenopausal women. Although the etiology of incontinence is multifactorial, it is likely that menopause plays a contributory role in the onset of symptoms of urinary incontinence and other urogenital disorders. Clinical investigations demonstrate that the amount of collagen in skin and urogenital tissue declines after menopause and that this deficit can be completely reversed by estrogen administration (Brincat et al. 1985). Furthermore, estrogens appear to increase the sensitivity of the a-adrenoreceptors to adrenergic stimulation, thus enhancing the contractile response of the urethral smooth muscle (Caine 1977). Compared with a placebo, estrogen therapy results in an overall significant subjective improvement in all types of incontinence (Fanth et al. 1994).

1.4.1.1 Women's Knowledge of and Attitudes towards HRT

Women from France, Germany, Spain, and the United Kingdom were interviewed to determine their knowledge of and views on menopause and HRT (Schneider 1997). Although there appears to be an increasing trend towards HRT usage, the fact remains that the majority of European women who might benefit from the therapy do not receive it. The fundamental reason for the low level of usage seems clear: Most of the women remain uninformed about the benefits of HRT. About two-thirds of respondents in this study believed they needed more information. Over 85% of women in France, Germany, and the United Kingdom and two-thirds of women in Spain spontaneously mentioned oral tablets when queried about available HRT formulations. Half of the women interviewed were not aware of the availability of transdermal HRT formulations, and fewer still were aware of gels, creams, or injections. Moreover, fewer than half of the women outside the United Kingdom viewed the long-term benefits of HRT – protection against both osteoporosis and cardiovascular disease – as very important. Only about one-quarter of the relatively small proportion of women currently using HRT predicted life-long HRT use, despite the fact that the full benefits of HRT may accrue only with long-term use.

A study of nearly 500 women in Germany found that just 25% of current users had been on HRT for more than five years (Schneider and Dören 1996). Over 50% of lapsed users in our cross-national study of HRT in Europe had taken HRT for up to two years, a result similar to that of a German study in which 43% of past HRT users had taken the medications for up to two years (Schneider and Dören 1996) and a British study where 51% of lapsed users stopped HRT use within one year (Hope et al. 1995). These findings suggest that a large percentage of women use HRT until menopausal symptoms abate, but choose not to continue the regimen over the long term.

It is apparent that increased physician–patient communication and public education programs are needed. In our European investigation, fewer than half of premenopausal or postmenopausal women and only about two-thirds of perimenopausal women had discussed menopause or its symptoms with either a primary-care physician or a gynecologist. Although over 80% of women in each country expected to be the final arbiters of whether to use HRT or not, and of which formulation to use, most women reported that the advice of a physician should be taken into account. Individual physicians can and should play a direct role by initiating discussions on HRT and menopause, by educating their patients on the issues, and by helping with the HRT decision-making process (Ferguson et al. 1989). In addition, physicians, their professional groups, and public health agencies need to promote the development of public education programs that inform women about the physiology of menopause and the general reasons for using HRT (Dören and Schneider 1996).

Individual physicians and health agencies should also become more sensitive to the wide variety of factors that can affect whether physicians prescribe HRT and whether patients accept the therapy. For example, a physician's gender may account, in part, for the variability in HRT-prescribing practices. A study of over 200 women in the United States found that those cared for by female physicians were found to be significantly more likely to receive HRT than women cared for by male physicians (55% vs 23%, respectively; $p<0.001$) (Seto et al. 1996). It is unclear if this reflects a more prevention-oriented practice style in female physicians, or if there is a higher level of comfort between patients and physicians of the same gender in talking about menopause, or if there are other structural or interpersonal factors. In addition, seemingly unrelated patient characteristics may affect women's behavior with respect to HRT use. For instance, in Germany, women who used oral contraceptives in the past were found to be substantially more likely to use HRT than women who had never used oral contraceptives (Schneider and Dören 1996).

1.4.1.2 Long-Term Therapy

Determining what their female patients know and feel about menopause and HRT may also help physicians to work with women in choosing their therapeutic regimen and to approve long-term compliance with the regimen selected. Physicians should discuss the advantages and disadvantages of alternative formulations with their patients and should describe formulations, such as transdermal estrogen patches, that may not be well known. For example, while both transdermal and oral HRT methods provide relief from vasomotor and urogenital symptoms, prevent loss of bone mass, and have beneficial effects on cardiovascular function, the transdermal route may have advantages with respect to bioavailability, patient satisfaction, and compliance. The transdermal route reduces estrogen metabolism by the liver and provides estrogen levels closer to those of premenopausal women. Moreover, studies have found that among women switched from oral to transdermal estrogen, 70% to 80% preferred the patch (Balfour and McTavish 1992). For women who do not wish to use transdermal estrogen or in whom patches result in side effects such as skin irritation, continuous combined oral estrogen/progesterone therapy might be preferable to sequential therapy, because patients using the no-bleed preparation have been found to be more compliant in the long term (Eiken and Kolthoff 1995).

1.4.2 Chronic Disease

1.4.2.1 Osteoporosis

At present, estrogens remain the standard in terms of prevention of osteoporosis and treatment of established osteoporosis in postmenopausal women. Estrogens produce multisystem effects with a fairly well-defined optimal bone-sparing dose (Table 1). Several epidemiological studies have shown that HRT provides protection against fractures (Weiss et al. 1980; Kiel et al. 1987). However, it is obvious that the treatment has to be continued for some years to be effective. A definite time period for treatment is difficult to establish, but once estrogens are prescribed in the immediate postmenopausal period, between five and ten years may be necessary before protection against osteoporotic fracture is obtained.

There is a variety of alternative treatment options for established osteoporosis (Table 2). Adequate calcium and vitamin D intake is essential for the development and maintenance of a normal skeleton. The likelihood of falling increases with advancing age, and efforts should be directed at protection against falling and its consequences (Grisso et al. 1991). Physical activity has a protective effect against the risk of hip fracture (Jaglal et al. 1993).

While estrogen prevents postmenopausal bone loss and, unequivocally, reduces fractures of both trabecular and cortical bone, other agents have proven to be of varying clinical importance (Table 3). Calcitonin (intranasal and injectable), an inhibitor of bone resorption, can also be used in the prevention of osteoporosis. Bisphosphonates have under-

Table 1. Dose of ERT required to prevent bone loss

	Spine	Hip
Premarin	0.625 mg	0.625 mg
Estrone sulfate	0.625 mg	1.25 mg
Estrace	0.5–2 mg	2 mg
17β estradiol	0.05 mg	0.05 mg
Ethinyl estradiol	7–10 µg	unknown

Table 2. Prevention of bone loss after menopause

Therapy	Efficacy
Estrogens	17β-estradiol CEE Estrone
Bisphosphonates	Alendronate
SERMs	Raloxifene
Calcitonin (nasal)	+/–[a]
Calcium/vitamin D	No
Phytoestrogens	Unknown

[a]An effect and no effect are both observed.

Table 3. Management of osteoporosis

Therapy[a]	Decreases fracture rates
Estrogen (CEE)	Yes[b,c]
Alendronate	Yes[b,c]
Calcitonin	Yes[b]
Raloxifene	Yes
Calcium	No[d]
Fluoride	Yes/No[c]

[a]All therapies include calcium supplementation.
[b]FDA-approved indication.
[c]Dose effect.
[d]Except for patients with low prior calcium intake.

gone a revolutionary development in recent years and are more than just the next generation of osteoporosis treatment. There are many different bisphosphonates, and each one has a different profile of activity. The most potent compounds contain a nitrogen atom within a heterocyclic ring. Recent work suggests that nitrogen-containing bisphosphonates inhibit enzymes of the mevalonate pathway, which is required for the modification (prenylation) of important signaling proteins. The absence of protein prenylation disrupts osteoclast function and inhibits bone resorption. New bisphosphonates, initially pamidronate and, more recently, alendronate and residronate make it possible to normalize the biochemical indices in a majority of patients. In this way it is, for example, possible to effectively suppress Paget's disease, a localized disorder of bone remodeling, characterized by increased osteoclastic bone resorption and secondarily increased osteoblastic new bone formation, in a mild stage, by a single 60 or 90 mg infusion of pamidronate;

the more severe disease requires multiple infusions. Alendronate, 40 mg daily for six months, also normalizes the indices in 60% of moderately to severely affected patients and oral residronate, 30 mg daily for only two months, restores indices to normal in 70% of moderately affected individuals (E.S. Siris 1998, personal communication). Residronate, a novel pyridinyl bisphosphonate, has been shown to bring about clinically significant improvements in patients with corticosteroid-induced osteoporosis (D.M. Reid 1998, personal communication). Other clinical indications for bisphosphonates, besides prevention of bone loss and treatment of osteoporotic bone fractures, include rheumatoid arthritis, other joint diseases, periodontal diseases and loosening of joint prostheses, osteogenesis imperfecta, and breast cancer bone metastases (for review, see Fleisch 1997).

Although much has been learned about sodium fluoride therapy for osteoporosis in the last twenty years, some questions still remain open. These concern the specific effect of fluorides on cortical and trabecular bone, variation in individual responsiveness and its efficacy in reducing vertebral fracture rates.

Tibolone, tamoxifen, and raloxifene have also been demonstrated to be efficient in the prevention of postmenopausal bone loss. The effect of raloxifene on BMD in postmenopausal women was examined in three large randomized, placebo-controlled, and double-blind osteoporosis prevention trials in North America, Europe, and in an international collaboration. The majority of the investigated women were Caucasian; the mean T scores for these three studies range from –1.01 to –0.74 for spine BMD, and included women with both normal and low BMD. Compared to calcium supplementation alone, raloxifene, at 60 mg once daily, produced increases in bone mass (Dexa measurements of hip, spine, and total-body BMD) (Delmas et al. 1997).

1.4.2.2 Cardiovascular Disease

Epidemiological studies conducted in northern Europe and the United States found that the cardiovascular protection is greater than the protection against or the risk of all other estrogen-associated conditions combined (Barrett-Connor 1997). On these grounds, HRT can be defended as a standard of care in a country such as the United States, where heart disease is the leading cause of death in postmenopausal women, and contributes significantly to female morbidity, with breast cancer rates much lower than heart disease rates. On the basis of meta-analyses of studies in which most women were treated with unopposed conjugated equine estrogen, HRT is associated with a 35% reduction in the risk of developing coronary heart disease; a healthy woman at no particular risk for heart disease, osteoporosis, or breast cancer would gain one year of life (Barrett-Connor 1997). Today's concept of primary prevention of cardiovascular disease by estrogens is universally accepted on the grounds of direct vascular and endothelial as well as metabolic actions.

Critical physicians are worried that the cardiovascular benefit is exaggerated, because women who use estrogen are more educated and healthier. If cardio-protection is exaggerated by various prevention, compliance, and prescription biases, what is the true risk reduction? The answer to this problem would depend on competing causes of morbidity and mortality in each country. This uncertainty has led to the initiation of various clinical trials, among them the secondary prevention trial, hormone estrogen–progestogen replacement study (HERS) (Herrington et al. 1998). While risks of secondary events increased in the first months of the study, they were reduced by about one-third within four to five years of HRT (Barrett-Connor 1997). Human coronary angiography studies, together with the long-term risk reduction found in the HERS trial suggest that HRT is beneficial, also in pre-existing cardiovascular disease.

1.4.2.3 Cognitive Function

During the reproductive life of women there is a more pronounced cerebral blood circulation than in men; this relation levels off after menopause. Clinical experience so far has demonstrated that estrogens improve cerebral vascularization after menopause; it is for this reason that estrogens, independent of their direct cardiovascular effect, play an important role in the prevention and treatment of vascular dementia.

About 20 investigations worldwide documented the influence of estrogens on the incidence of stroke in aging women. These studies were essentially corrected for risks such as age, systolic blood pressure, diabetes, and smoking habits. In those investigations with more appropriate controls, the stroke incidence, after 5–10 years of estrogen replacement, was reduced by approximately 20%.

Cognitive function in aging women was investigated in the design of a prospective study of women who underwent surgical menopause;

Sherwin (1994) could demonstrate that women on estrogens (and/or androgens) performed better in several tests of memory and logical reasoning than women taking placebo. In older women, study results are conflicting.

In women with cerebrovascular disease (transient ischemic attacks and/or reversible ischemic neurological deficits), the use of estrogens increased cerebral blood flow and resulted in better cognitive scores compared to the baseline, whereas the results in control groups became worse (Funk et al. 1991).

With increasing longevity, senile dementia has become a major health problem, with an impact on the quality of life of not only the patients, but also their care providers. After the age of 65, the prevalence of dementia and Alzheimer's disease doubles every five years, so that as many as 50% of women aged 85 years and older may suffer from this disability.

Estrogen has been shown to protect neurons from oxidative cell death in vitro, possibly through receptor-independent antioxidant activity; it also modulates neuronal survival and repair by stimulating nerve growth factor activity. Estrogens may affect neuronal function by modulating either neurotransmitter systems or glucose utilization in neurons. Estrogens also reduce serum levels of the amyloid precursor apolipoprotein E, which accelerates deposition of β-protein in the senile plaques of Alzheimer's disease. The preventive effect of estrogens on the clinical manifestation of Alzheimer's disease has been documented in several studies. In particular, Tang et al. (1996), in their community-based study of 11,024 women, reported estrogen use in 12.5% of these women. The age of Alzheimer's disease onset was significantly later in women who had taken estrogen compared to those who had not. The relative risk of the disease was 5.8% in estrogen users versus 16.3% in non-users. Kawas et al. (1997), reporting on a prospective study of 472 post-menopausal women followed up for 16 years, indicated that the risk of Alzheimer's disease was reduced to 0.64 in women who had reported the use of estrogen.

1.4.2.4 Contraindications and Risks of HRT

Absolute contraindications to HRT include breast cancer, active thrombosis, pregnancy, active liver disease, and undiagnosed vaginal bleeding. Several relative contraindications, including a history of endometrial cancer or pulmonary embolism, should also be considered when making decisions concerning HRT use. HRT is relatively contraindicated in women treated for endometrial carcinoma, unless the treating oncologist agrees with therapy and the patient is informed about its risks and benefits. In fact, studies have found significantly lower relapse rates for stage-I endometrial carcinoma in patients receiving HRT than in patients not receiving therapy (Creasman 1991).

The most recent review of the epidemiological studies on postmenopausal hormone replacement and the risk of breast cancer was presented in a collaborative group analysis (Collaborative Group on Hormonal Factors in Breast Cancer 1997). The individual data of 52,705 women with breast cancer and 108,411 women without breast cancer, from 51 studies in 21 countries, were analyzed. Among current users of HRT and all those who ceased use one to four years previously, the relative risk of having cancer diagnosed increased by a factor of 1.023 for each year of use. However, this increase in risk is comparable with the effect of delaying menopause, since for women who never used HRT, the relative risk of developing breast cancer increased by a factor of 1.028 for each year older at menopause. We expect that the data over the next decade will be obtained from controlled experimental trials rather than being based on observations. So far, any excess risk of breast cancer following HRT should be viewed as a possibility but not as a certainty.

Cigarette smoking adversely affects estrogen metabolism and is linked with early natural menopause. Smokers on HRT, however, have been reported to have the same or lower rates of myocardial infarction than nonsmokers on estrogen. Although patients should be encouraged to stop smoking, those who are unable or unwilling to do so may still be candidates for HRT.

1.5 Figures on HRT Use

Our international market analysis includes the European, North American, Latin American, and a composite of various African, Asian, and Australian countries (Table 4). To arrive at a basis for calculating the rate of HRT acceptance, we proceeded as follows:

Table 4. Countries included in our international market analysis

Europe	**North America**
Germany	USA
United Kingdom	Canada
France	
Italy	Latin America
Sweden	Brazil
Belgium	Mexico
Poland	Colombia
Spain	Chile
Austria	Argentina
Switzerland	Venezuela
Finland	Peru
The Netherlands	Central America
Norway	Ecuador
Turkey	Uruguay
Denmark	Dominican Republic
Czech Republic	
Portugal	**Africa/Asia/Australia**
Ireland	Australia
Slovakia	Japan
Greece	South Africa
Russian Federation	New Zealand
	Pakistan
	Thailand

- We determined the number of packs sold for the treatment of climacteric complaints.
- The number of treatment cycle units were calculated from the number of packs.
- Division of the treatment cycle units by 13 gives an estimate of the number of users over one year.
- The number of users are then related to all women aged 45 to 64.
- The resultant percentage of women represents the rate of acceptance.

An estimate of the growth in the climacteric female population worldwide and by continent is summarized in Fig. 1. Over the 20 years from 1990 to 2010, the population of females aged 45 to 64 years is expected to have doubled in Latin America, increased by about two-thirds in the United States and Canada, one-third in Europe, and about 50% in the

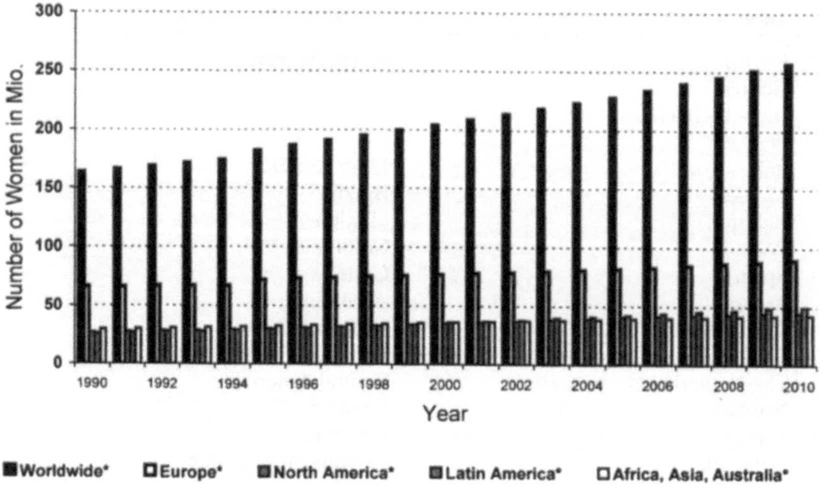

Fig. 1. Estimate of female population aged 45 to 64 years (from WDI/World Bank 1997; Schering Deutschland GmbH, Market Research, T. Hein; Schering AG, International Market Research)

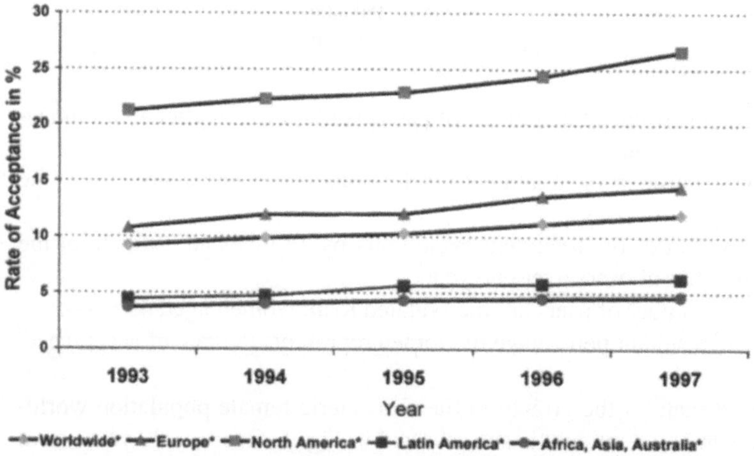

Fig. 2. Acceptance of HRT among women aged 45 to 64 (from WDI/World Bank 1997; IMS Health, Midas; Schering Deutschland GmbH, Market Research, T. Hein; Schering AG, International Market Research)

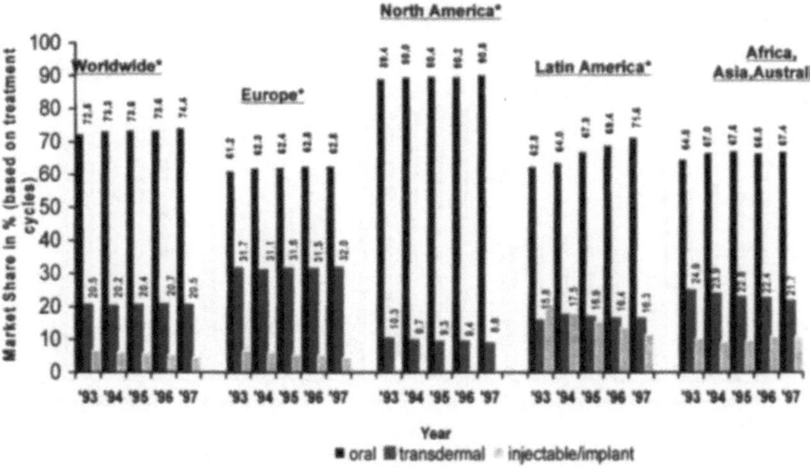

Fig. 3. Modes of application of HRT (from WDI/World Bank 1997; IMS Health, Midas; Schering Deutschland GmbH, Market Research, T. Hein; Schering AG, International Market Research)

Africa, Asia, and Australia taken as a whole. It is also expected to have increased by about 50% worldwide.

Acceptance of HRT varies widely. Although the acceptance worldwide has increased by more than 25% from 1993 to 1997, this trend mainly reflects habits in the United States and Europe. Markets in Latin America and Africa, Asia, and Australia have grown slowly from a rather low starting point (Fig. 2).

With respect to the mode of application, the oral route clearly dominates in the United States and Canada, with about 90%, which persisted through the five-year period of 1993 to 1997; likewise, transdermal application tended to stay at approximately 10% and less. In Europe, over the same period of time, there is a consistent two-third preference for the oral over the one-third of transdermal HRT. In Latin America, the oral route of administration increased from 63% to 72%, while the transdermal application stayed rather constant between 16% and 17.5%. Figures of Africa, Asia, and Australia list oral application constantly at around two-thirds of HRT takers and transdermal application decreased from one in four to one in five. The worldwide figures stabilized at three

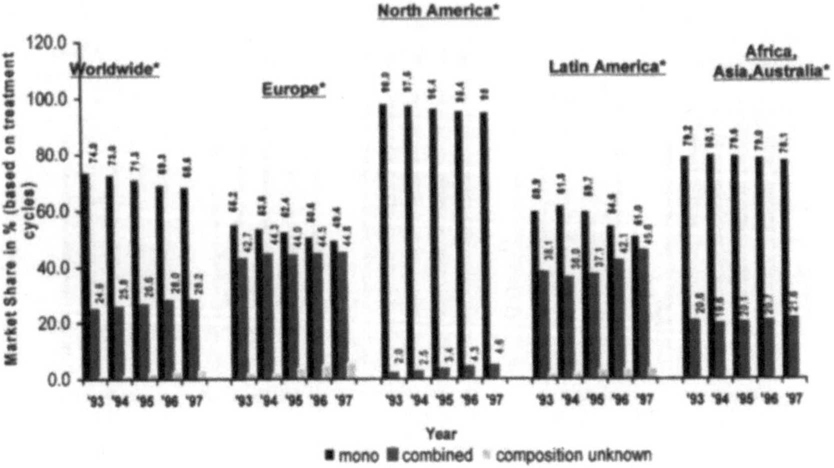

Fig. 4. HRT therapeutic regimens (from WDI/World Bank 1997; IMS Health, Midas; Schering Deutschland GmbH, Market Research, T. Hein; Schering AG, International Market Research)

out of four women on oral HRT, and one in five on transdermal HRT (Fig. 3).

Thus, the oral route is still the most common mode of administration. Estrogens are, however, available in various formulations for different routes such as intramuscular with estradiol esters, subcutaneous with estradiol implants, percutaneous with estradiol cream, transdermal with estradiol patch, intranasal with estradiol spray, sublingual with estradiol tablets, and vaginal with creams, tablets, and a silastic ring. These additional modes have been tabulated as "others" and have achieved major importance in Latin America and Africa, Asia, and Australia, but are less frequently used in Europe and are of extremely low impact in the United States and Canada.

Any HRT regimen can differ in estrogen and progestogen dosage, route of administration, and duration of use. There are two basic HRT regimens, each with modifications: unopposed estrogen therapy (ERT) and estrogen combined with progestogen (HRT). In the latter, there is the tradition of the sequential mode, with continuous estrogen and either the sequential or cyclic addition of a progestogen for twelve to fourteen days in order to avoid endometrial hyperplasia. Currently, a monthly regimen is recommended as the method of choice for nonhysterectomized women. Discontinuous administration of estrogens does not offer any advantages. The more recent approach is continuously combined HRT with its purpose to produce amenorrhea whilst achieving sufficiently high serum estrogen levels; this has gained great importance in bleeding control and improved compliance. Our worldwide search in the mid-1990s did not allow for looking at more than mono- and combined therapeutic regimens, leaving out a proportion of unknown compositions (Fig. 4). Worldwide there is a decreasing tendency to use ERT, from 74% to 68.6%, whereas the use of combined regimens increased from 24.9% to 28.2%. There are, however, large variations in the various continents. The tendency in Europe is clearly in favor of combined HRT, starting from a 12.5% dominance of monotherapy to almost equity with 49.4% monotherapy compared to 44.8% combination therapy in 1997. A similar trend is seen in Latin America, while Africa, Asia, and Australia apparently stabilized at around 80% mono- and a little over 20% combination therapy.

In a country such as Germany, which has a long tradition of use of the sequential and continuous-combined mode of HRT, figures are as depicted in Figs. 5 and 6. In 1997, there was a predominance of the sequential (46.3%) and continuous modes (11.6%) of application compared to monotherapy (42.1%). A look into the German market, on the other hand, also allows for differentiation between the progestogens (Fig. 7). This market is dominated by norethisterone, medrogestone, and levonorgestrel, while use of cyproterone acetate and desogestrel range between 2.5 and 6%.

We are certainly aware of the problems involved in a proper interpretation of market and over-the-counter sales data as indicators of hormone use in the population. The rate of acceptance of HRT calculated for Germany according to the above-mentioned standard was 13.1% in 1998 and shifted via 22.5% in 1991 to 31.1% and 31.8% in 1994 and 1997, respectively. On the other hand, our representative study of nearly 500 women in Germany found that 24% of women were current users of HRT and that 25% of these had been on HRT for more than five years (Schneider and Dören 1996). The difference between 31% HRT acceptance over the counter and 24% admittedly ingested by peri- and postmenopausal women aged 45 to 60 years is only partly explained by

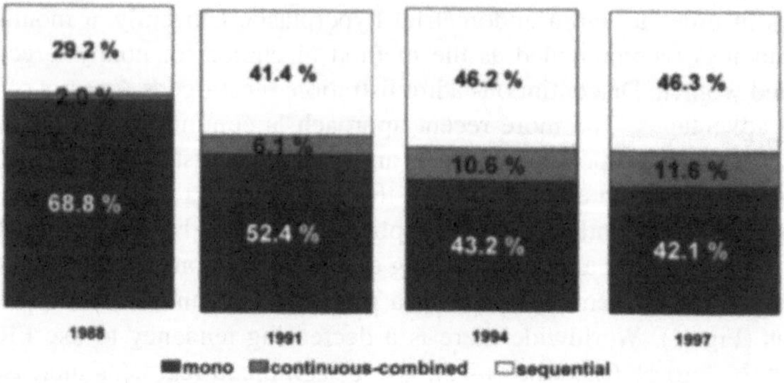

Fig. 5. HRT therapeutic regimens in Germany (from IMS Health, Institute for Medical Statistics, Pharmacy Market Germany; Schering Deutschland GmbH, Market Research, T. Hein, 18 Aug 1998)

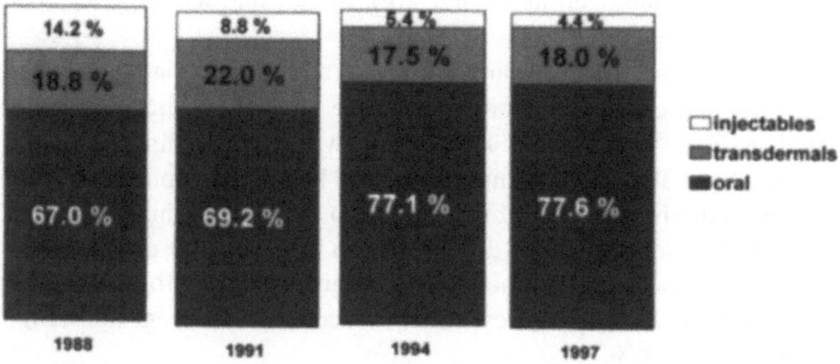

Fig. 6. Modes of application of HRT in Germany (from IMS Health, Institute for Medical Statistics, Pharmacy Market Germany; Schering Deutschland GmbH, Market Research, T. Hein, 18 Aug 1998)

looking at a certain age segment; the main difference points to the proportion of women who hand in their prescription but do not take the drug.

HRT market figures are very difficult to get access to. Table 5 allows some insight into the major markets of the United States and Europe. These countries in themselves roughly constitute a US $3 billion enter-

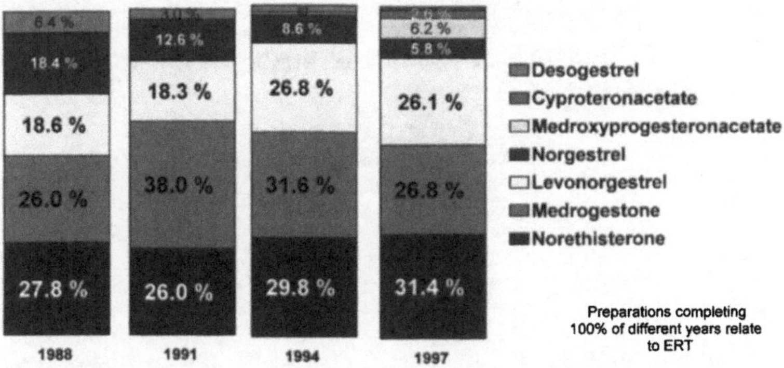

Fig. 7. Market share of different progestogens in German HRT prescriptions (from IMS Health, Institute for Medical Statistics, Pharmacy Market Germany; Schering Deutschland GmbH, Market Research, T. Hein, 18 Aug 1998)

Table 5. HRT market worldwide (in thousands of US dollars)

	1993[a]	1994[a]	1995[a]	1996[a]	1997[a]	2/1998[b]
USA						
E only	805,513	885,229	949,721	1,043,631	1,114,645	1,169,134
E and P	–	–	25,364	157,155	302,570	362,474
Europe						
E only	272,462	300,174	355,048	393,108	384,538	398,158
E and P	230,183	281,717	365,187	413,340	432,145	444,491
UK						
E only	45,480	53,006	60,162	68,690	79,029	81,242
E and P	75,409	86,496	103,884	119,500	138,225	145,645
Germany						
E only	73,535	76,408	83,631	81,122	74,013	70,830
E and P	94,115	109,757	136,530	136,407	128,244	123,361
France						
E only	53,939	55,500	72,289	79,597	79,185	80,507
E and P	9,182	15,706	20,009	21,648	22,032	22,847

E, estrogen; P, progestogen.
[a]Year/End 4th quarter.
[b]Moving annual total of 2nd quarter.

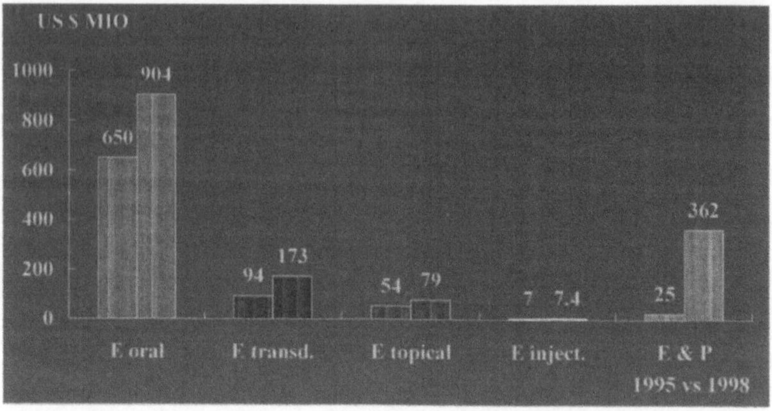

Fig. 8. HRT market – Mode of application in the United States 1993 vs 1998

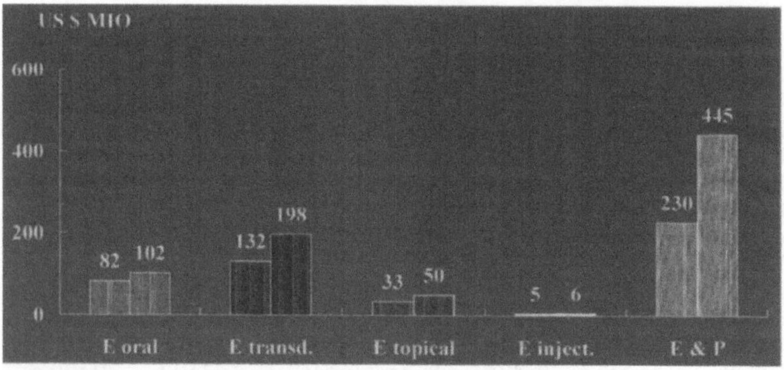

Fig. 9. HRT market – Mode of application in Europe 1993 vs 1998

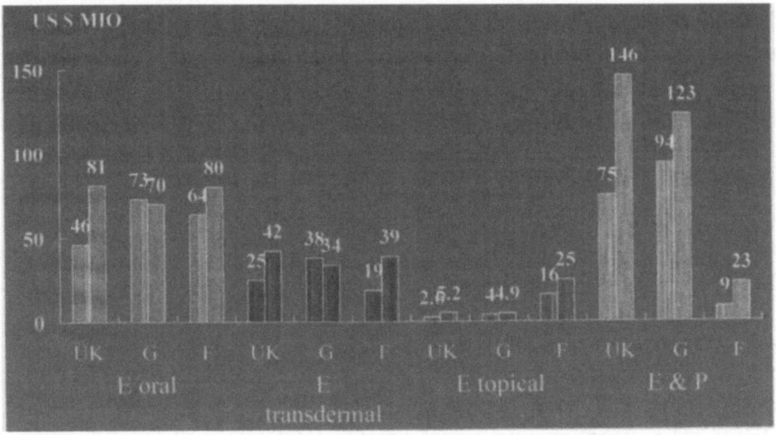

Fig. 10. HRT market – Mode of application in the United Kingdom, Germany, and France, 1993 vs 1998

prise, with the United States clearly in the lead over Europe; within Europe, it is the United Kingdom that takes the lead with a quarter billion US dollar share, closely followed by Germany and, a little more distant, France. While estrogen monotherapy is still gaining in the US market, the US market has, since the mid-1990s, grown 15-fold with respect to estrogen and progestogen combinations. In Europe, the combination therapy has taken the lead ever since 1995 and has had more than a 10% lead in 1998. This difference in favor of estradiol and progestin preparations is even more clearly seen in the United Kingdom and Germany, where combination therapy almost reaches double the proportion of estrogen monotherapy. Surprisingly enough, in France, ever since 1996, estrogen-only preparations sell almost fourfold more than combination products.

A more detailed look at modes of application is illustrated in Figs. 8, 9, and 10. A six-year trend (1993 vs 1998) documents the persisting dominance of oral estrogen administration in the United States over transdermal and topical routes as well as local injections. The combination therapy, as already mentioned, is gaining drastically. In Europe, transdermal therapy remained the leading route of estrogen application, with an increasing trend; the estradiol and progestogen combination therapy, however, has stayed on top of the sales since 1993 and has almost doubled during this period (Fig. 9). A direct comparison of modes of application as interpreted from market shares (Fig. 10), besides showing a trend in favor of oral estrogens in the United Kingdom, demonstrates larger gains in the transdermal routes in the United Kingdom and France in contrast to Germany, and a dominance of topical application in France compared to the other two countries. As far as the addition of a progestogen is concerned, all countries follow similar positive trends although from rather different baselines.

Worldwide figures convincingly demonstrate that standard HRT, besides estrogen monotherapy, incorporates combination oral or transdermal estrogens. Although the benefits of long-term use of HRT are widely recognized, most women use HRT for relief of climacteric complaints. The most important variable for HRT use is the continent or country involved, despite the fact that countries investigated are close, both culturally and geographically. The data on worldwide HRT acceptance, its modes of application and routes of administration suggest that there are important differences in habits and accepted opinion regarding the treatment of the climacteric and its sequelae. As seen from our cross-national study of women's use of HRT in Europe (Schneider 1997), decisions about beginning HRT and choosing a formulation are viewed by most women as matters of personal choice, to be made with advice from a physician. Despite the benefits of HRT and available choices among drug-delivery options, a fairly small proportion of women use it, largely because most women remain fully uninformed about the therapy. Physician–patient communication and public education certainly represent keys to HRT acceptance and judicious decisions.

1.6 Future Aspects

Many physicians worldwide think that there is sufficient evidence to recommend estrogen as the standard of care for the postmenopausal woman, on the basis of the consistent if circumstantial evidence from observational studies and the number of positive estrogen effects on lipoproteins, coagulation factors, and coronary artery endothelium and smooth muscle. Others argue that benefits such as those to the cardiovascular system may be exaggerated, because women who use estrogen are more educated and healthier. The Women's Health Initiative (WHI) is a primary prevention trial of fairly healthy women (although women with heart disease are not excluded). It will include 27,000 women assigned either the same regimen as that used in HERS, or unopposed estrogen for women who have had a hysterectomy. Recruitment for this study should be completed in 1997 and the study will continue for 9 more years. Patients and physicians are polarized with regard to the need for compliance with longstanding or lifetime HRT. Some physicians are recommending that their peri- and postmenopausal patients who have no immediate indications for HRT should wait 10 years for the intervention trial results before making an estrogen decision. Their argument is that these women will be unlikely to suffer a heart attack or hip fracture, since over 80% of these events occur after age 70. Others, persuaded by the consistency and biological plausibility of the data, do not think that clinical trials are indispensable and even point to placebo-controlled long-term investigations as unethical (Barrett-Connor 1997). In the meantime, current investigations and clinical experience with tissue-selective estrogens and alternatives to estrogen and progestogen replacement such as bisphosphonates, phytohormones, and SERMs are very promising; it is very unlikely that the medical world would hamper clinical application of such newly developed drugs until decades of intervention trials have passed.

The presentation of available information on current HRT use worldwide should help to assist physicians and their patients to follow global experience. On the other hand, I would like to refer to the French philosopher Voltaire and his view of dealing with the process of aging: "Qui n'a pas l'esprit de son âge, de son âge a tout le malheur."

Acknowledgments. To arrive at serious demographic estimates of the peri- and postmenopausal female population worldwide, and in order to quantify the acceptance of HRT, its modes of application, and therapeutic regimens, we were supported by Schering Deutschland GmbH, Market Research; T. Hein, Schering AG, International Market Research; and Wyeth Corporation Germany.

References

Balfour JA, McTavish D (1992) Transdermal estradiol: a review of its pharmacological profile, and therapeutic potential in the prevention of postmenopausal osteoporosis. Drugs Aging 2:487–507

Barrett-Connor E (1997) Menopause: Problems and interventions in the United States. In: Paoletti R, Crosignani PG, Kenemans P, Samsioe G, Soma M, Jackson AS (eds) Women's health and menopause. Kluwer, Dordrecht, pp 9–13

Beresford SAA, Weiss NS, Voigt LF, McKnight B (1997) Risk of endometrial cancer in relation to use of estrogen combined with cyclic progestogen therapy in postmenopausal women. Lancet 359:458–461

Brincat M, Moniz CJ, Studd JWW, Darby A, Magos A, Emburey G, Versi E (1985) Long-term effects of the menopause and sex hormones on skin thickness. Br J Obstet Gynaecol 92:256–259

Caine M (1977) The importance of adrenergic receptors in disorders of micturition. Eur Urol 3:1–6

Collaborative Group on Hormonal Factors in Breast Cancer (1997) Breast cancer and hormone replacement therapy: collaborative reanalysis of data from 51 epidemiological studies of 52705 women with breast cancer and 108411 women without breast cancer. Lancet 350:1047–1059

Creasman WT (1991) Estrogen replacement therapy: is previously treated cancer a contra-indication? Obstet Gynecol 77:308–312

Delmas PD, Bjarnason NH, Mitlak BH, Ravoux AC, Shah AS, Huster WJ, Draper M, Christiansen C (1997) The effects of raloxifene on bone mineral density, serum cholesterol, and uterine endometrium. N Engl J Med 337:1641–1647

Dören M, Reuther G, Minne HW, Schneider HPG (1995) Superior compliance and efficacy of continuous combined oral estrogen-progestogen replacement therapy in postmenopausal women. Am J Obstet Gynecol 173:1446–1451

Dören M, Schneider HPG (1996) The impact of different HRT regimens on compliance. Int J Fertil 41:29–39

Eiken P, Kolthoff N (1995) Compliance with long-term oral hormonal replacement therapy. Maturitas 22:97–103

Fanth JA, Cardozo L, McClish D (1994) Estrogen therapy in the management of urinary incontinence in postmenopausal women: a metaanalysis. Obstet Gynecol 83:12

Ferguson KJ, Hoegh C, Johnson S (1989) Estrogen replacement therapy: a survey of women's knowledge and attitudes. Arch Intern Med 149:133–136

Fleisch H (1997) Bisphosphonates in bone disease, 3rd edn. Parthenon, New York, pp 1–184

Fries JF (1980) Aging, natural death and the compression of morbidity. N Engl J Med 303:130
Fries JF (1988) Aging, illness, and health policy: implications of the compression of morbidity. Perspect Biol Med 31:407
Fries JF, Green LW, Levine S (1989) Health promotion and the compression of morbidity. Lancet 1:481
Funk JL, Mortel KF, Meyer JS (1991) Effects of estrogen replacement therapy on cerebral perfusion and cognition among postmenopausal women. Dementia 2:268–272
Grisso JA, Kelsey JL, Strom BL et al. (1991) Risk factors for falls as a cause of hip fracture in women. N Engl J Med 324:326–331
Herrington DM, Fong J, Sempos CT et al. (1998) Comparison of the HERS cohort with women with coronary heart disease from the NHANES III Survey. Am Heart J 136:115–124
Hope S, Rees MCP, Radcliffe J (1995) Why do British women start and stop hormone replacement therapy? J Br Menopause Soc 12:16–17
Jaglal SB, Kreiger N, Darlington G (1993) Past and present physical activity and risk of hip fracture. Am J Epidemiol 138:107–118
Kawas C, Resnick S, Morrison A, Brookmeyer R, Corrada M, Zonderman A, Bacal C, Lingle DD, Metter E (1997) A prospective study of estrogen replacement therapy and the risk of developing Alzheimer's disease: the Baltimore Longitudinal Study of Aging. Neurology 48:1517–1521
Kiel DP, Felson DT, Andersen JJ, Wilson PWF, Moskowitz MA (1987) Hip fracture and the use of estrogens in postmenopausal women. The Framingham study. N Engl J Med 317:1169–1174
Persson I, Adami H-O, Bergquist L et al. (1989) Risk of endometrial cancer after treatment with oestrogens alone or in conjunction with progestogens: results of a prospective survey. Br Med J 298:147–151
Rodin J (1986) Aging and health: effects of the sense of control. Science 233:1271
Schneider HPG (1997) Cross-national study of women's use of hormone replacement therapy (HRT) in Europe. Int J Fertil 42 (Suppl 2):365–375
Schneider HPG, Dören M (1996) Traits for long-term acceptance of hormone replacement therapy – results of a representative German survey. Eur Menopause J 3 (Suppl 2):94–98
Seto TB, Taira DA, Davis RB, Safran C, Phillips RS (1996) Effect of physician gender on the prescription of estrogen replacement therapy. J Gen Intern Med 11:197–203
Sherwin BB (1994) Estrogenic effects on memory in women. Ann NY Acad Sci 734:213–230

Tang MX, Jacobs D, Stern Y, Marder K, Schofield P, Gurland B, Andrews H, Mayeux R (1996) Effect of estrogen during menopause on risk and age at onset of Alzheimer's disease. Lancet 348:429–432

Weiss NS, Ure CL, Ballard JH, Williams AR, Dalin JR (1980) Decreased risk of fractures of the hip and lower forearm with postmenopausal use of estrogen. N Engl J Med 303:1195–1198

2 General Aspects of HRT in Japan

T. Aono, K. Azuma, M. Irahara

In the 1960s, estrogen replacement therapy (ERT) was introduced in the United States, and many postmenopausal women underwent this therapy, seeking a better physical condition, the so-called "feminine forever". In the late 1970s, the incidence of endometrial cancer was found to be several times higher among ERT users than in non-users. After this report, the number of ERT users decreased. In the early 1980s, combined administration of progestogen with estrogen was found to be effective for reducing endometrial cancer, and since then hormone replacement therapy (HRT) with estrogen and progestogen has been widely used for postmenopausal women. Recently, 10%–30% of postmenopausal women in North American and European countries have been undergoing HRT, however, the proportion of HRT users among Japanese women aged 45–60 years is only 1%.

In this chapter, we will report on general aspects of HRT in Japan, and then analyze why most Japanese women do not use this effective treatment for hypoestrogenic conditions.

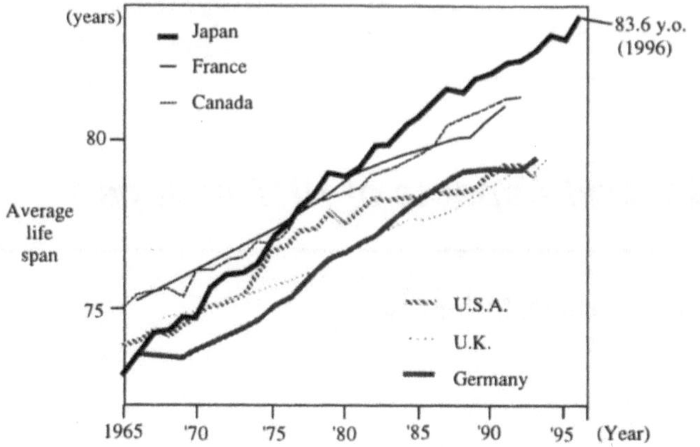

Fig. 1. International comparison of the average female life span (Health and Welfare Statistics Association 1998)

2.1 The Average Life Span and the Age of Menopause

The average life span of women is continuously increasing in Western countries and in Japan, as shown in Fig. 1 (Health and Welfare Statistics Association 1998a). The average life span of Japanese women in 1996 reached 83.6 years, and this is the longest in the world.

Khaw reported that the average age of menopause is about 50 years, and is constant in all eras and races (Khaw 1992). Tamada and Iwasaki studied the age of menopause in 456 Japanese women. They reported that the age of menopause varies from the early 40s to the mid-50s, and that its average is 49.47±3.526 years (Tamada and Iwasaki 1995). This study on the average age of menopause in Japanese women agrees with Khaw's theory.

From these data, that the average life span is getting longer and that the age of menopause is constant, the number of postmenopausal Japanese women can be estimated. The Statistics Bureau of Management and Coordination Agency estimated that the number of Japanese postmenopausal women continuously increase, as shown in Fig. 2 (Statistics Bureau of Management and Coordination Agency 1993). The population of Japanese postmenopausal women in 1990 was 20.48 million, and it is estimated that it will be 26.19 million in the year 2000 and 29.07

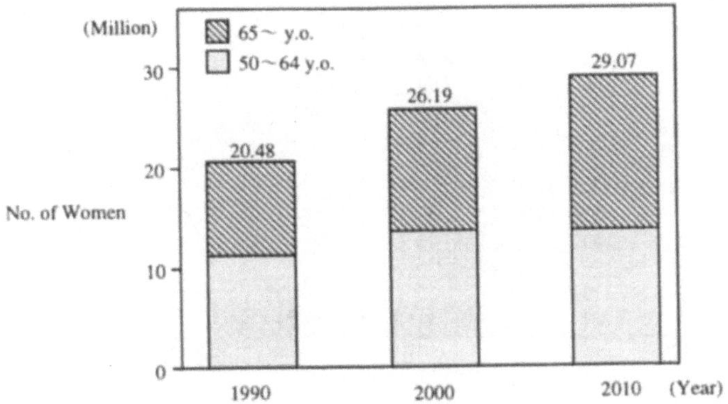

Fig. 2. The estimated population of Japanese women after menopause (Statistics Bureau of Management and Coordination Agency 1993)

million in 2010. These figures show that the number of postmenopausal women will increase by 50%, that is, by as many as 9 million, in 20 years.

2.2 The Proportion of Elderly People

The total Japanese population in 1997 was 126,166,000. It is estimated that the total Japanese population will reach a peak (127,780,000) in 2008, and that it will decrease thereafter to 100,496,000 in 2050 (National Institute of Population and Social Security Research 1998). These figures show that the total population in Japan will decrease by 20% in the next 50 years.

How will the proportions of elderly people in Japan in the next century look like, when it is considered that the total Japanese population will decrease and the number of postmenopausal Japanese women will increase?

The age distribution of the Japanese population in 1996 and 2050 is shown in Fig. 3 (Ministry of Health and Welfare 1998a). In 1996, the largest population group comprises people aged 40–49, but in 2050, people aged 70–80 is estimated to make up the largest group. Furthermore, the proportion of people aged 65 years or more was 14.5% in

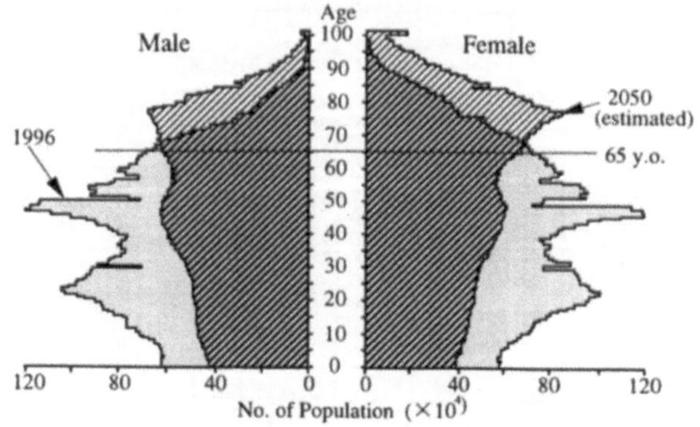

Fig. 3. The age distribution of the Japanese population (Ministry of Health and Welfare 1998)

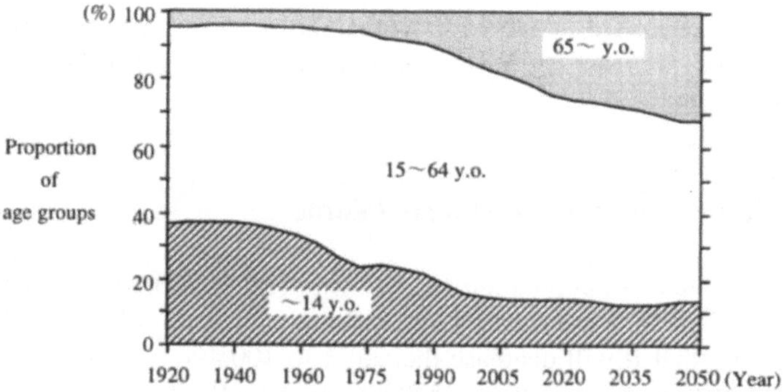

Fig. 4. The changes in the proportions of the age groups in the Japanese population (Ministry of Health and Welfare 1998)

1995, and will be 32.3% in 2050, as shown in Fig. 4 (Ministry of Health and Welfare 1998b). This means that one-third of the Japanese population will be 65 years old or more. An international comparison of the proportion of people aged 65 years or more is shown in Fig. 5 (Ministry of Health and Welfare 1998c). In the next century, Japan will have the highest proportion of elderly people in the world, and will become an "aging society".

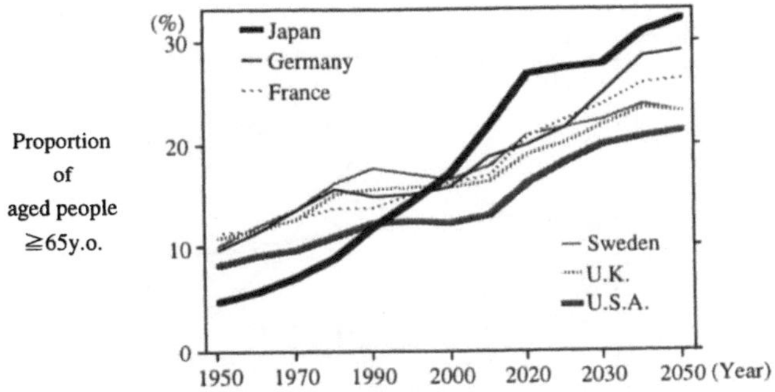

Fig. 5. International comparison of the proportion of aged people (³65 years old) (Ministry of Health and Welfare 1998)

If this situation is taken in consideration, then the question of how we can maintain elderly people's health will be one of the most important issues in the next century.

2.3 Disease and HRT

Before the Second World War, tuberculosis had the highest mortality rate in Japan. The treatment of infectious diseases was dramatically improved by the introduction of antibiotics, and the mortality rates of pneumonia and tuberculosis decreased in these 50 years. However, neoplasm, heart disease, and cerebral apoplexy still have high mortality rates, as shown in Fig. 6 (Health and Welfare Statistics Association 1998b).

Many reports show that the frequency of endometrial cancer decreases by HRT, however, the frequency of breast cancer is considered to slightly increase under the influence of HRT.

The Japanese Ministry of Health and Welfare reported that the number of bedridden old people was about 900,000 in 1993, and it is estimated that 1,200,000 people will be bedridden in 2000 (Ministry of Health and Welfare 1998d).

Several diseases cause old Japanese people to be bedridden, as shown in Fig. 7 (Minaguchi 1997). Cerebral apoplexy, hypertension, decrepi-

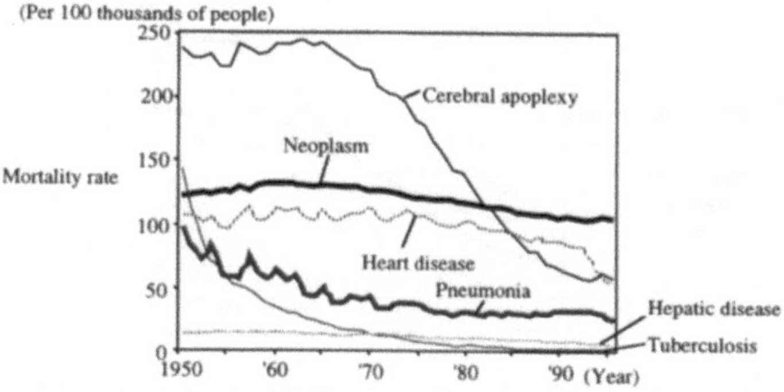

Fig. 6. The changes in the mortality rates due to the major diseases in Japanese women (Health and Welfare Statistics Association 1998)

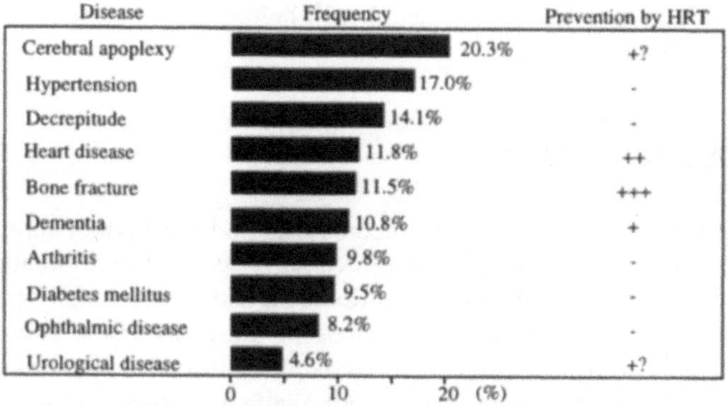

Fig. 7. Diseases causing old Japanese people to be bedridden and the possibility of their prevention by HRT (modified from Minaguchi 1997)

tude, heart disease, and bone fracture are diseases with frequencies above 10%.

Many bone fractures of old people are related to osteoporosis, and significant numbers of bone fractures can be prevented by HRT. Furthermore, some prospective studies reported that the odds ratio of the risk of ischemic heart disease and cerebral vascular disease can be reduced by HRT (Meade and Berra 1992). Ohkura also reported that some women with dementia of the Alzheimer type may be treated with HRT (Ohkura et al. 1994). These data show that the increase in bedridden Japanese women, which will be one of the most serious issues in the next century, may be prevented to some extent by HRT .

2.4 The Prevalence of HRT Use in Japan

Several kinds of tablets containing estrogen and a patch containing estradiol are available for HRT in Japan. They consist of conjugated equine estrogen (Premarin, Asahi-Kasei), estradiol (Estraderm TTS, Novartis), estriol (Estriel, Mochida), and estrone, pregnenolone, androstenedione, testosterone, and thyroid hormone (Metharmone-F, Nihon-Zouki). The sales of these drugs are shown in Fig. 8. Since 1990, their sales have been increasing gradually. However, the proportion of Japanese women aged 45–64 who use HRT is about 1% or less, and is much lower than in Western countries, where 10%–30% of women use HRT (Fig. 9).

It is not clear why most Japanese women do not use HRT; however, we tried to analyze this problem.

Firstly, many Japanese women and physicians think that "aging" itself is a natural process, and that it should not be disturbed by artificial intervention. In a sense, old Japanese women are accepting their aging process as their mothers and grandmothers did.

Secondly, most older Japanese women are not in the habit of taking hormonal tablets daily, because they did not use oral contraceptives

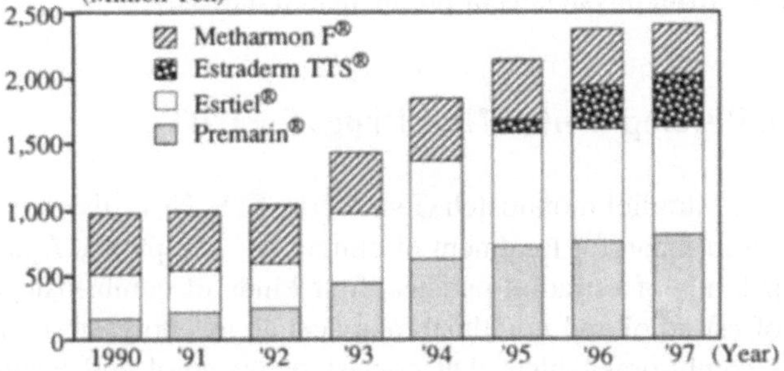

Fig. 8. Medication for HRT prescribed in Japan (IMS Japan. K.K.)

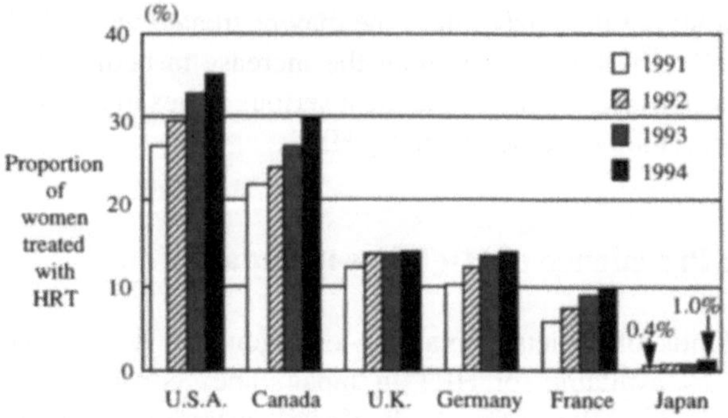

Fig. 9. The prevalence of HRT in women (45–64 years old) (IMS Japan. K.K.)

(OCs) while they were young. In many Western countries, 70%–80% of young women use OCs. They take this sort of tablet every day for years. In contrast, 70%–90% of Japanese women use barrier methods for contraception, and the proportion of Japanese women who are OC users is less than 2%. Most Japanese women are therefore not familiar with hormonal medication.

Thirdly, Japanese women are afraid of cancers and the side effects of HRT. The information on HRT provided by physicians or public health nurses is not enough. The Japanese Ministry of Health and Welfare performed a questionnaire survey on climacterium. The question was: "From whom did you get information about climacterium?" Japanese women got information from (1) mass media, (2) friends, and (3)mothers and sisters. It may be that most Japanese women do not get enough information from physicians or public health nurses.

2.5 The Development of New Drugs for HRT

In 1995, an estradiol monopatch (Estraderm TTS, Novartis) was put on the market in Japan for treatment of climacteric symptoms. Apart from this, four kinds of estradiol patches, four kinds of combi-patches that consist of estradiol and norethisterone acetate levonorgestrel, and two kinds of combi-oral tablets that consist of estradiol and levonorgestrel/trimegestone are undergoing clinical trials.

As drugs for osteoporosis, three kinds of estradiol patches, a combi-patch with estradiol and norethisterone acetate, and two kinds of combi-oral tablets that consist of estradiol and levonorgestrel/trimegestone are being developed. Furthermore, tibolone and raloxifen are synthetic oral steroid hormones that are also available for osteoporosis; they are undergoing phase III of clinical trials.

2.6 Hormone Treatment Strategy for Women

Recently, several clinical trials were performed on low-dose oral contraceptive use in premenopausal women with hypoestrogenic symptoms. Trossarelli and colleagues (Trossarelli et al. 1995) administered low-dose oral contraceptives to 58 women aged 35–49 years (mean age 40.2 years) to treat climacteric symptoms, such as hot flushes, sweating,

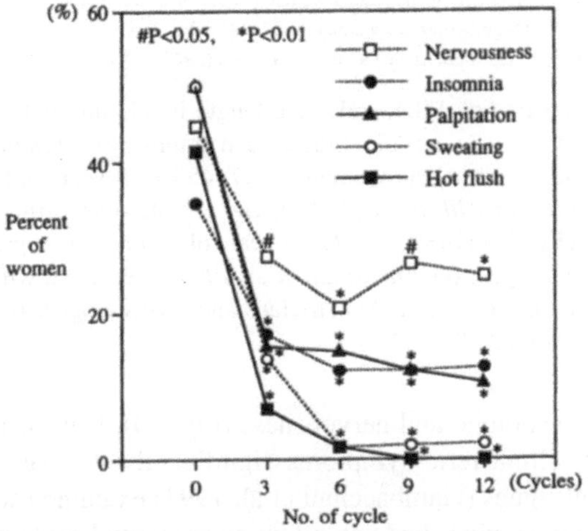

Fig. 10. The effects of the low-dose pill on climacteric symptoms. Subjects: 58 women aged 35–49 years (mean age: 40.2 years). OC: ethinyl estradiol 20 μg+desogestrel 0.15 mg (Reprinted from Contraception, 51, Trossarelli GF, Gennarelli G, Benedetto C et al., Climacteric syndromes and control of the cycle in women aged 35 years or older taking an oral contraceptive with 0.150 mg desogestrel and 0.020 mg ethinyl estradiol, pp 13–18, 1995, with permission from Elsevier Science)

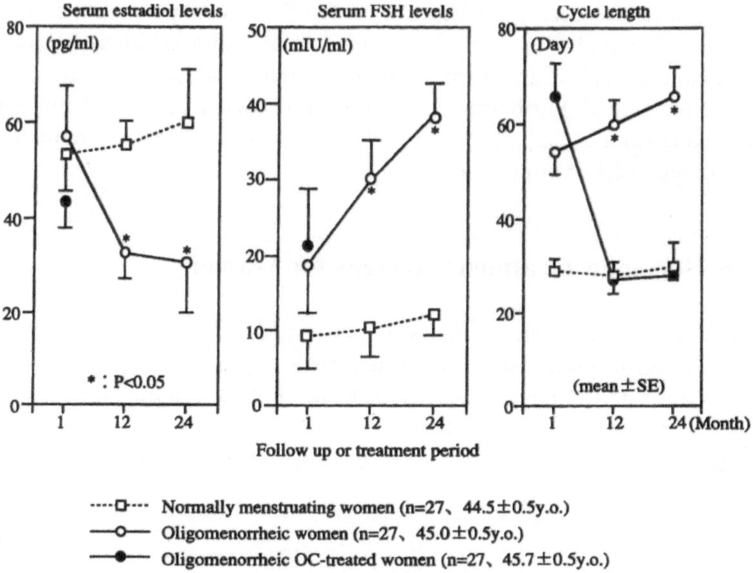

Fig. 11. Serum estradiol, FSH, and cycle length in: (1) normally menstruating women (n=27, 44.5±0.5 years old; indicated in graphs by *broken lines with open squares*); (2) oligomenorrheic women (n=27, 45.0±0.5 years old; indicated in graphs by *solid lines with open circles*), and (3) oligomenorrheic OC-treated perimenopausal women (n=27, 45.7±0.5 years old; indicated in graphs by *solid lines with filled circles*) (Gambacciani et al. 1994). Reprinted with permission from the American College of Obstetricians and Gynecologists (Obstetrics and Gynecology, 1994, 83, 392–396))

palpitations, insomnia, and nervousness (Fig. 10). During 12 treatment cycles, these climacteric symptoms significantly decreased. Gambacciani and colleagues (Gambacciani et al. 1994) examined whether low-dose oral contraceptives had any effect on menstrual cycle irregularities (Fig. 11). They gave OCs to three groups of perimenopausal women. The first group consisted of 27 women (44.5±0.5 years) with normal menstruation cycles. The second group consisted of 27 women (45.0±0.5 years) with oligomenorrhea, and the third group consisted of 27 women (45.7±0.5 years) with oligomenorrhea and treated with OCs. In the group of oligomenorrheic women, serum estradiol levels decreased, serum FSH levels increased, and the cycle length prolonged

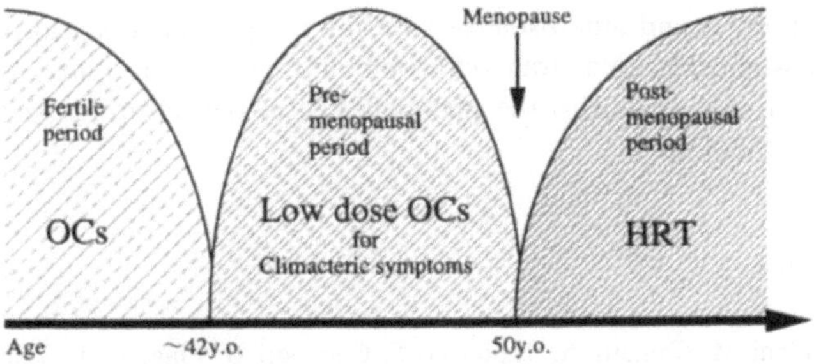

Fig. 12. Hormonal treatment strategy for women to deal with contraception and hypoestrogenic disorders

over 24 months. In the group of oligomenorrheic women treated with OCs, cycle lengths were normalized to 28.0±1.0 days.

From these data, it is concluded that the low-dose contraceptive pills are effective as hormonal supplementation for premenopausal women with climacteric symptoms or menstrual irregularity.

Generally speaking, women wish to use contraception during their fertile period, and wish treatment of climacteric symptoms during the premenopausal period. After menopause, hypoestrogenic disorders, osteoporosis, and atherosclerosis need to be prevented. A strategy for the hormonal treatment of women can be designed to span the fertile period to the postmenopausal period, with OCs, low-dose OCs, and HRT (Fig. 12).

2.7 Conclusion

Japan is now approaching a new era in which one-third of its women will be 65 years or older. We must be prepared to cope with this unprecedented situation. HRT is proven to be effective against the uncomfortable symptoms experienced during climacterium, and to be effective against a substantial proportion of osteoporosis cases. Therefore, HRT should be adopted as one of the major strategies for coping with an "aging society" in Japan.

Women want contraception during their fertile period, treatment of climacteric symptoms during the premenopausal period, and prevention of osteoporosis and atherosclerosis during the postmenopausal period. Women who undergo a strategic hormonal treatment designed to span from the fertile period to the postmenopausal period are expected to spend happier lives.

References

Gambacciani M, Spinetti A, Taponeco F, Cappagli B, Piaggesi L, Fioretti P (1994) Longitudinal evaluation of perimenopausal vertebral bone loss: effects of a low-dose oral contraceptive preparation on bone mineral density and metabolism. Obstet Gynecol 83:392–396

Health and Welfare Statistics Association (1998a) The life tables for Japan. J Health Welfare Stat 45:76–80

Health and Welfare Statistics Association (1998b) Death in Japan. J Health Welfare Stat 45:48–60

Khaw KT (1992) Epidemiology of menopause. Br Med Bull. 48:249

Meade TW, Berra A (1992). Hormone replacement therapy and cardiovascular disease. Br Med Bull 48:276–308

Minaguchi H (1997) Osteoporosis in 10 million people. Shufu-no-tomo, Tokyo, p 25

Ministry of Health and Welfare (1998a) The population pyramid of Japan. In: Annual report on health and welfare. Gyousei, Tokyo, p 365

Ministry of Health and Welfare (1998b) The estimated age proportions of the Japanese population. In: Annual report on health and welfare. Gyousei, Tokyo, p 367

Ministry of Health and Welfare (1998c) The proportion of people aged 65 years or more in developed countries. In: Annual report on health and welfare. Gyousei, Tokyo, p 367

Ministry of Health and Welfare (1998d) Health and welfare for the elderly. J Health Welfare Stat 45:119–130

National Institute of Population and Social Security Research (1998) J Health Welfare Stat 45:37–42

Ohkura T, Isse K, Akazawa K et al. (1994) Evaluation of estrogen treatment in female patients with dementia of the Alzheimer type. Endocr J 41:361–371

Statistics Bureau of Management and Coordination Agency (1993) J Health Welfare Stat 40:52–62

Tamada T, Iwasaki H (1995) Age at natural menopause in Japanese women. Acta Obst Gynaec Jpn 47:947–952

Trossarelli GF, Gennarelli G, Benedetto C et al. (1995) Climacteric syndromes and control of the cycle in women aged 35 years or older taking an oral contraceptive with 0.150 mg desogestrel and 0.020 mg ethinyl estradiol. Contraception 51:13–18

3 Recent Advances in Steroid Receptor Research: Focusing on Estrogen Receptors

M. Muramatsu, S. Ogawa, T. Watanabe, K. Ikeda, H. Hiroi, A. Orimo, and S. Inoue

3.1 Introduction

Estrogen is mainly secreted from the ovary and plays an important role in the development and maintenance of the reproductive system in the female. It mainly exerts its effect by binding to the specific receptors present in each target cell. The mode of action of the estrogen receptor (ER) is illustrated in Fig. 1. The ER waits for the incoming estrogen (the most potent physiological one being 17β-estradiol, E_2) in the cell nucleus, and when this ligand comes into the cell through the cytoplasm,

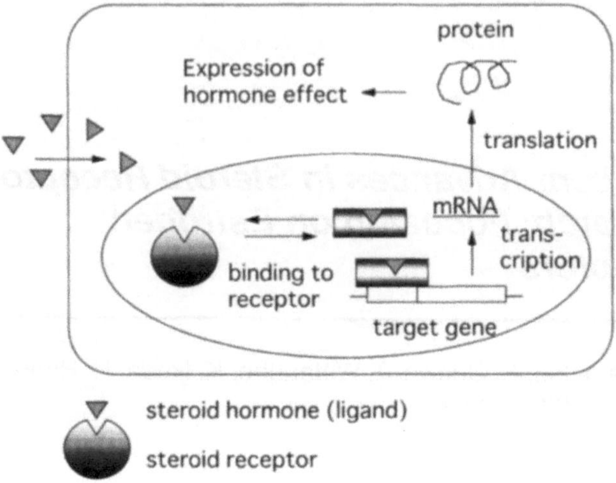

Fig. 1. Schematic diagram of estrogen action through an estrogen receptor

the ER binds E_2 with a very high affinity, and then binds to a certain DNA sequence, named the estrogen responsive element (ERE) and activates the specific genes that have this element (Jensen and DeSombre 1973, Evans 1988, Green and Chambon 1988). In this sense, ER may be said to be a ligand- (E_2-) dependent transcription factor.

Other steroid receptors are also known, including androgen receptors (AR), progesterone receptors (PR), glucocorticoid receptors (GR), and mineralocorticoid receptors (MR), each of which is activated by its specific hormonal ligand. There are other kinds of receptor molecules too, which bind to the thyroid hormone, to vitamin D_3, and to retinoic acids (vitamin A derivatives), and there are also other endogenous as well as exogenous ligands which regulate various aspects of cell differentiation and function in the same manner as steroid receptors do; they all together form the large superfamily of the nuclear receptors (Mangelsdorf et al. 1995).

In this article, we will focus on ER research, with special reference to the newly discovered ERβ, which is providing a new dimension in the understanding of estrogen action.

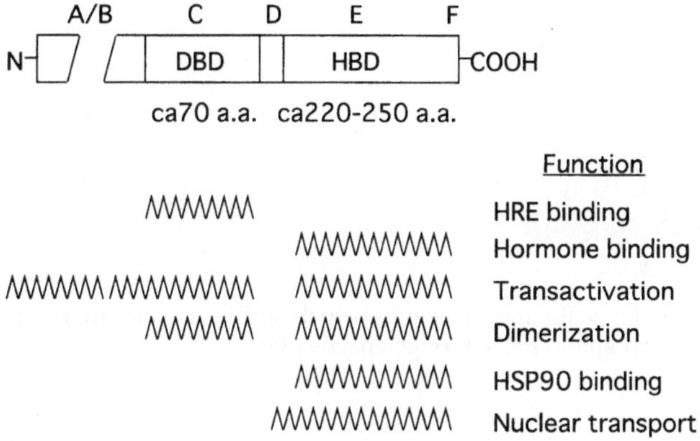

Fig. 2. Structure of a typical nuclear receptor. DBD: DNA-binding domain; HBD: hormone-binding domain

3.2 The Estrogen Receptor as a Transcription Factor. Identification of Cofactors

The structure of a typical nuclear receptor is depicted in Fig. 2. Generally speaking, there is a DNA-binding domain in the center and a hormone- (or ligand-) binding domain in the C-terminal region. As these molecules bind as dimers to the DNA, the dimerization domain is also in this region. For activation of the gene, there are two distinct domains at the N-terminal and the C-terminal regions. Of course, the presence of the DNA-binding domain is mandatory for transactivation. Just how these domains work on the preinitiation complex (PIC) on the target gene promoter remains to be determined. However, a new class of molecules, including SRC-1, TIF1, TIF2, AIB1, CBP/p300, etc., are now being demonstrated to act as coactivators by bridging between ER and PIC (Kamei et al. 1996, Chakravarti et al. 1996, Heery et al. 1997, Voegel et. al. 1998, Puigserver et al. 1998), and a scheme such as that illustrated in Fig. 3 is emerging. This interaction and activation via coactivators or mediators are now being investigated vigorously, to make up a most advanced model of transcription regulation. A detailed mechanism of estrogen (and other nuclear receptor) action will be

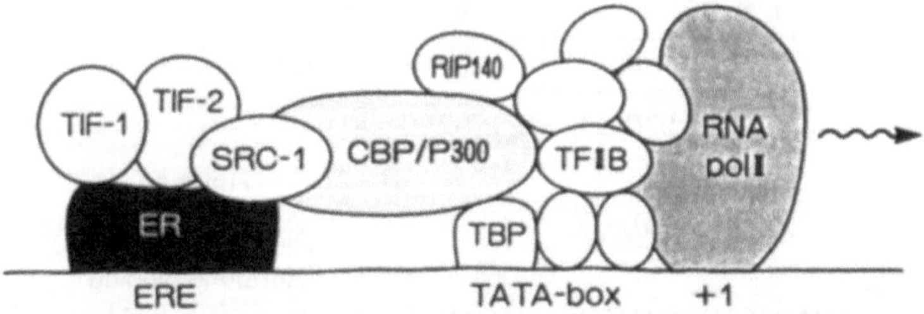

Fig. 3. A model of the mechanism by which an ER activates transcription at a preinitiation complex (PIC) through coactivators

clarified in the not so distant future, but will not be discussed in this chapter because of space limitations.

3.3 Discovery of ERβ, a Subtype of the ER Family

For a long time, estrogen action had been believed to be mediated by only one ER. There were, however, some doubts that there was only one receptor species, because the knockout mice of this gene (ERKO mice, Lubahn et al. 1993) did not completely lose their female characteristics, the presence of the uterus, mammary glands, and female genitalia, although they were hypoplastic and mostly nonfunctional. In 1996, Kuiper et al. (1996) discovered another subtype of ER in the rat and named it ERβ; thus the previously known ER was renamed ERα.

Human and mouse ERβs were also cloned by Mosselman et al. (1996) and Tremblay et al. (1997), respectively. All their amino acid sequences, however, lacked 53 amino acids at the N-terminus of the molecule. The first complete amino acid sequence of human ERβ (hERβ) was determined by our group (Ogawa et al. 1998a) and was found to contain 530 amino acids with M_r 59.2 kDa (Fig. 4). hERβ has an amino acid identity of 96% in the DNA-binding domain (C), relative

Fig. 4. Structure of hERβ cDNA, with the complete amino acid sequence of hERβ (**A**); also shown is a structural comparison with hERα (**B**). *Underlined* are the primers which were used for RT-RCR and the additional 53 amino acid sequence which was lacking in previous publications

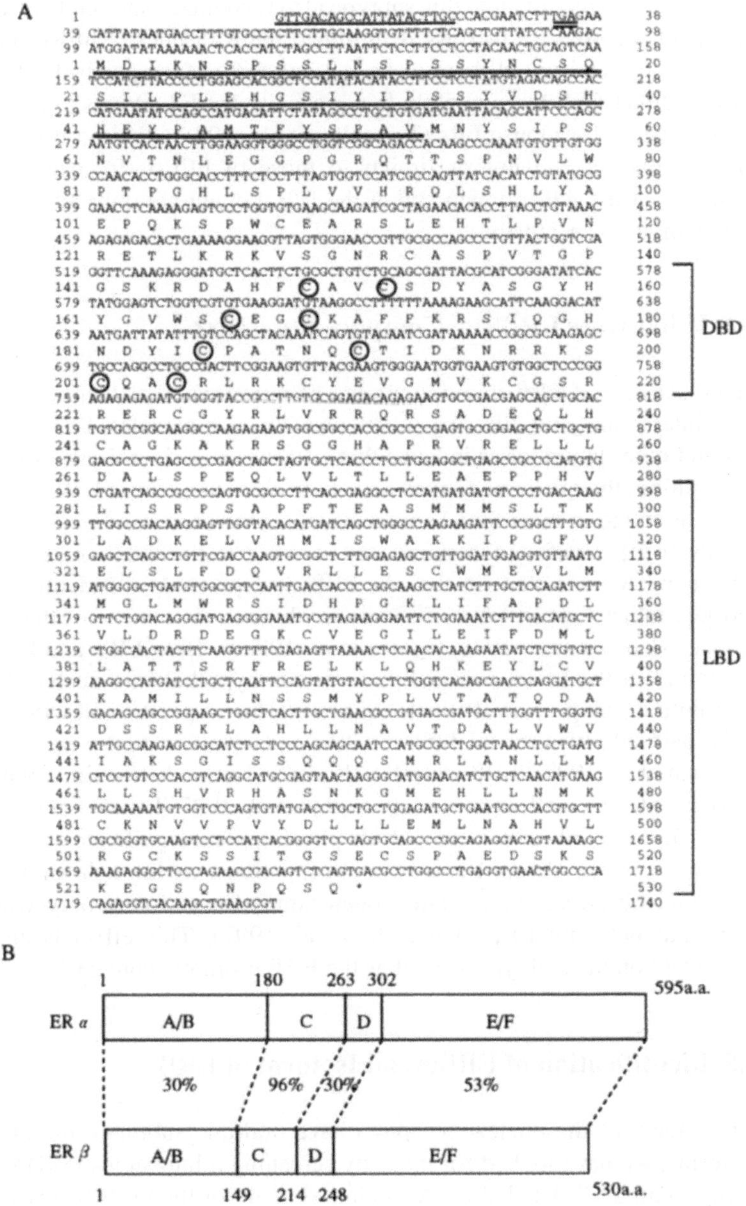
A
DBD
LBD
B
ER α
A/B
C
D
E/F
1
180
263
302
595a.a.
30%
96%
30%
53%
ER β
A/B
C
D
E/F
1
149
214
248
530a.a.

to hERα; this suggests that this subtype also recognizes and binds ERE, just as hERα does. In fact, we determined the K_d values of hERα and hERβ to E_2, and found them to be 0.2 and 0.6 nM, respectively; this indicates that ERβ has a slightly lower affinity for E2. Others reported similar values for mouse ERs (Tremblay et al. 1997). Other portions of hERβ, however, are not so similar to those of hERα; for example, identities at the A/B domain and at the E/F domain are only 30% and 53% similar, respectively.

3.4 What Does ERβ Do, Compared to ERα?

The discovery of ERβ raised a number of questions as to the distribution, interaction, and differing functions of these molecules. Studies in our and other laboratories indicate that ERβ has a distinct tissue or cell distribution, though some tissues have both receptors (Fig. 5). Whereas hERα is found in most female organs such as the uterus and mammary glands, hERβ was found in abundance in ovary, testis, and thymus. It is interesting to note that, in the rat, ERβ is expressed, abundantly, in prostate rather than in testis, a situation which is the reverse of that in humans (Kuiper et al. 1996, Ogawa et al. 1998b). Both ERα and ERβ are found in the brain, especially at various hypothalamic nuclei, the distribution being again distinct, suggesting their specific roles in the CNS, particularly in sexual differentiation and behavior (McEwen 1994, Rissman et al. 1997). ERβ was also shown to form a heterodimer with ERα in vitro as well as in vivo (Cowley et al. 1997, Pettersson et al. 1997, Ogawa et al. 1998a) though its physiological significance is not known at present. As for transactivation functions, we have shown that the agonistic effect of the antiestrogen tamoxifen is found only with ERα, but not with ERβ (Watanabe et al. 1997). This effect is also dependent on the cell type as well as the ERE promoter context.

3.5 Identification of ERβcx, an Isoform of ERβ

Since many of the nuclear receptors have multiple subtypes and also isoforms, we have looked for those by screening a human testis cDNA library with a ^{32}P-labeled DNA-binding domain of the rat ERα cDNA

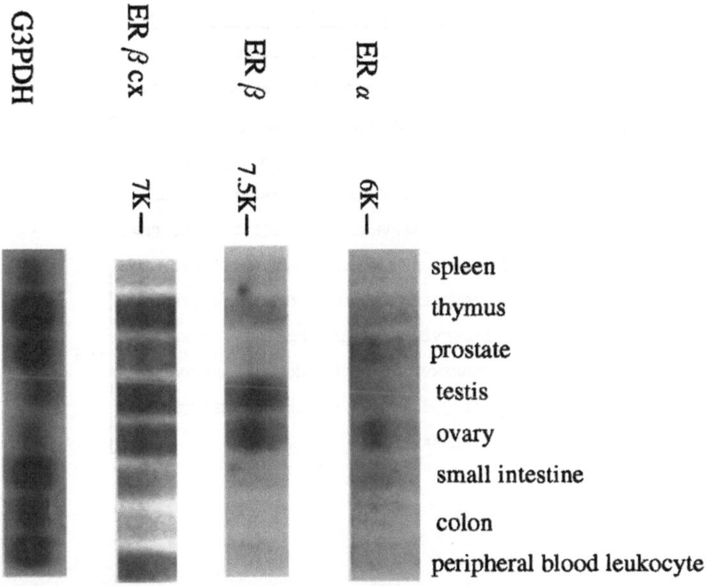

Fig. 5. Expression of ERα, ERβ, and ERβcx mRNA in different human tissues. Northern blots with poly(A)+ mRNA from various tissues were hybridized with ^{32}P-labeled cDNA probes which were specific for each ER mRNA. G3PDH was used as an internal control

(Ogawa et al. 1998a), and obtained a novel splice isoform designated hERβcx (Ogawa et al. 1998b). Interestingly, hERβcx cDNA was found to have a nucleotide sequence identical to that of hERβ up to the first 469 amino acids, but it had 26 unique amino acids instead of the C-terminal 61 amino acids of hERβ. We further obtained the genomic clones of the hERβ gene from a human genomic DNA library (Japanese Cancer Research Resources Bank) and analyzed the relationship between hERβ and hERβcx. As shown in Fig. 6A,B, hERβcx was found to be derived from a specific exon that was present 3' to the last ERβ exon, and alternative splicing produced either ERβ or ERβcx. The production of isoforms by this type of alternative splicing was already known for other nuclear receptors such as the human glucocorticoid receptor (hGR), the human thyroid hormone receptor (hTR) and rat vitamin D_3 receptor (rVDR) (Fig. 6B). It should be noted that, in all cases, the AF-2

A

```
  1  MDIKNSPSSLNSPSSYNCSQSILPLEHGSIYIPSSYVDSH
 41  HEYPAMTFYSPAVMNYSIPSNVTNLEGGPGRQTTSPNVLW
 81  PTPGHLSPLVVHRQLSHLYAEPQKSPWCEARSLEHTLPVN
121  RETLKRKVSGNRCASPVTGPGSKRDAHFCPVCSDYASGYH
161  YGVWSCEGCKAFFKRSIQGHNDYICPATNQCTIDKNRRKS   DBD
201  CQACRLRKCYEVGMVKCGSRRERCGYRLVRRQRSADEQLH
241  CAGKAKRSGGHAPRVRELLLDALSPEQLVLTLLEAEPPHV
281  LISRPSAPFTEASMMMSLTKLADKELVHMISWAKKIPGFV
321  ELSLFDQVRLLESCWMEVLMMGLMWRSIDHPGKLIFAPDL
361  VLDRDEGKCVEGILEIFDMLLATTSRFRELKLQHKEYLCV
401  KAMILLNSSMYPLVTATQDADSSRKLAHLLNAVTDALVWV   LBD
441  IAKSGISSQQQSMRLANLLMLLSHVRHAR/AEKASQTLTSF
481  GMKMETLLPEATMEQ*     (496 a.a )
```

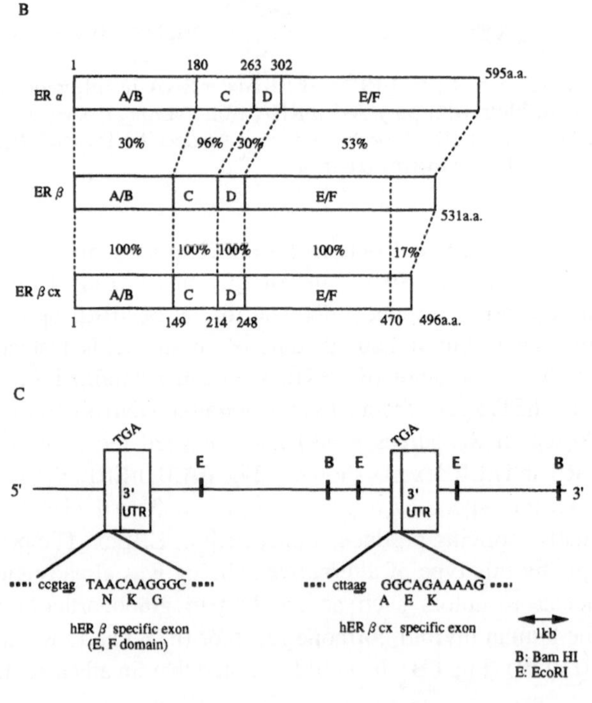

(activation function 2) core sequence is removed by the alternative splicing, leading to the loss of the AF-2 function, as shown below. The hERβcx mRNA was expressed in human testis, ovary, prostate, and thymus (Fig. 5). In contrast, ERα mRNA was expressed in ovary and prostate, but ERβ mRNA was seen in testis and ovary in human, as mentioned before. The presence of ERβcx protein is confirmed by Western blotting (data not shown, cf., Ogawa et al. 1998b). The meaning of this differential expression remains to be elucidated.

3.6 ERβcx May Regulate E_2 Action Negatively in Humans

To test the function of ERβcx, various experiments were performed. First, the binding of [^{3}H]-17β-estradiol to this receptor was tested and found to be almost null in transfected COS-7 cell extracts, while, as mentioned before, ERα and ERβ bound well with this ligand (Ogawa et al. 1998b). Next, the binding of these ERs to a ^{32}P-labeled consensus ERE (from a *Xenopus vitellogenine* gene) was examined by gel-shift assay. As shown in Fig. 7A, ERα and ERβ each produced a specific band which could be competed for by nonlabeled ERE, whereas ERα could not produce any shifted band. Then, we studied the effect of ERβcx on the binding of ERα to the ERE. The results shown in Fig. 7B clearly demonstrate that the presence of ERβcx in the cell extract dose-dependently inhibits the binding of ERα to the ERE; this suggests that ERβcx may, in some way, interfere with the function of ERαβ.

We therefore examined the functional aspects of these ERs by a chloramphenicol acetyltransferase (CAT) assay. Figure 8A shows that ERα strongly enhances the activity (25-fold of control) of the construct made up of ERE–globin promoter–CAT (ERE–GCAT). ERβ also has significant enhancing activity, albeit a little lower (9-fold) under these conditions. Note that these activities are totally E_2-dependent. ERβcx, however, has no enhancing activity whatsoever, as expected from its inability to bind to ERE (see Fig. 7A). It has been demonstrated that a

◄

Fig. 6. Genomic, organization of hERβ exons involved in producing hERα and hERβcx (**A**), similar alternative splicing found in other nuclear receptors (**B**), and deleted C-terminal sequences of the variants that contain the AF-2 core region (*boxed*, **C**). Splice junctions are shown by *perpendicular bars* in **C**

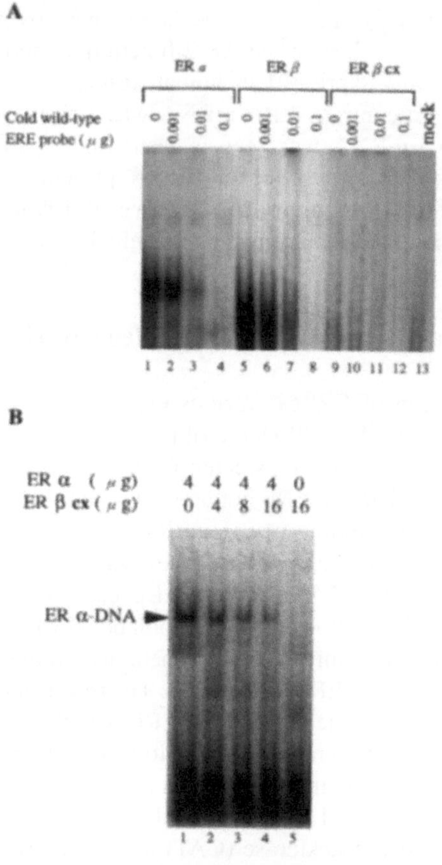

Fig. 7A,B. DNA-binding abilities of ERα, ERβ and ERβcx. Whole cell extracts of COS-7 cells transfected with the indicated ER expression vectors were analyzed by gel-shift assay with ^{32}P-labeled ERE as probe. Increasing amounts of cold ERE were used for competition (**A**). Binding of ERα to ERE was competed for by increasing amounts of ERβcx (**B**)

A

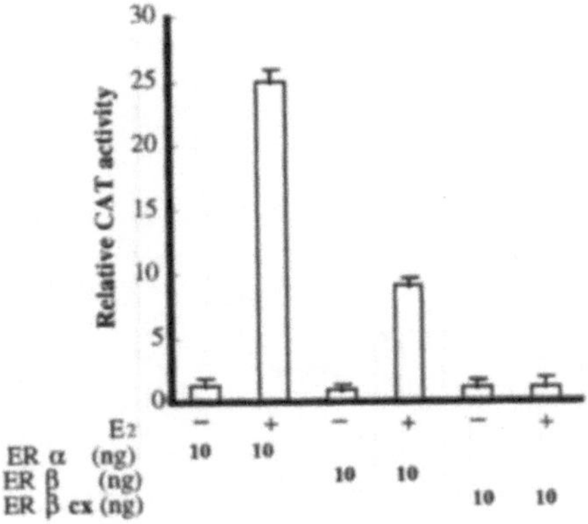

B

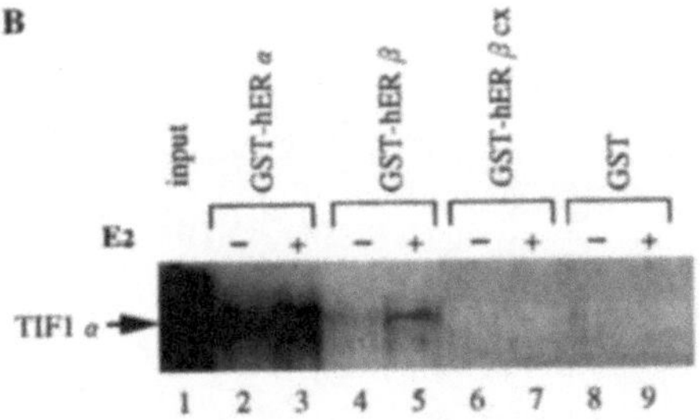

Fig. 8A,B. Transcriptional enhancing activity of ERα, ERβ, and ERβcx and their binding ability to a coactivator TIF-lα. Transactivation of the ERE–globin promoter–CAT reporter gene by ERα, ERβ or ERβcX expression vector (**A**). The lack of interaction between ERβcx and TIF-1α (**B**). GST pull-down assay was performed as described

transcriptional coactivator TIF-1 interacts with the AF-2 domain of ERα and activates gene transcription by remodeling the chromatin structure (Le Douvarin et al. 1996). We therefore examined the interaction of ERβcx with TIF-1; for this we used a glutathione-S-transferase (GST) pull-down assay. The results shown in Fig. 8B clearly indicate that both ERα and, to a lesser extent, ERβ bind TIF-1α, whereas ERβcx cannot bind it under the same conditions (Ogawa et al. 1998b).

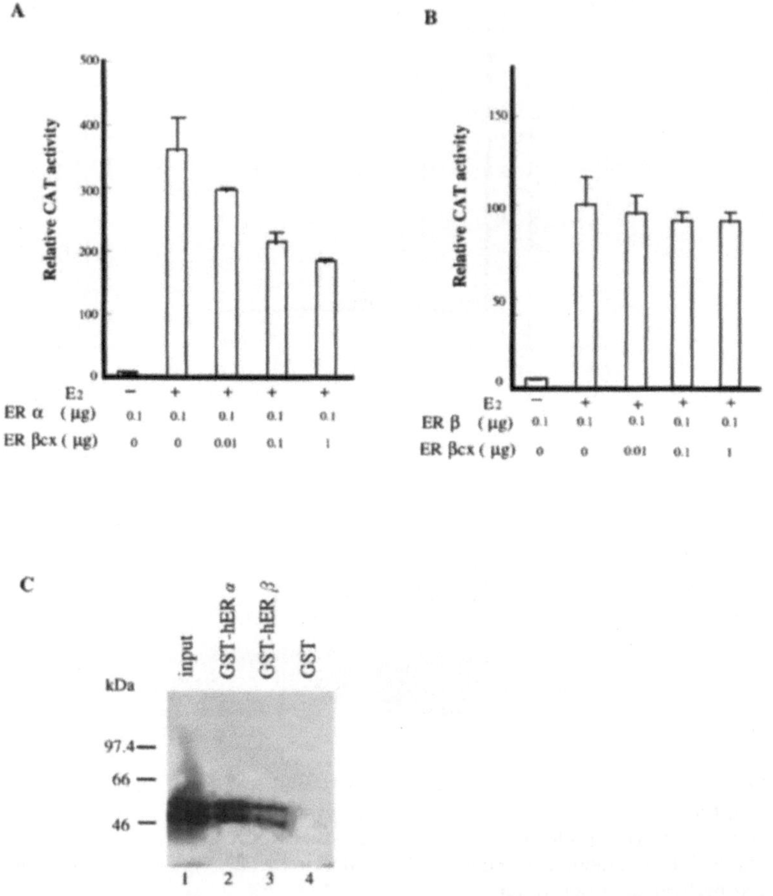

Fig. 9A-C. Dominant negative activity of ERβcx in the ERα signaling pathway. Dose-dependent inhibition of ERα transactivation by ERβcxα (**A**). The abscissa represents the amount of transfected DNA (expression vector) and the ordinate shows the CAT activities of the reporter. Little inhibition is shown for ERβ by ERβcx (**B**). The GST pull-down assay shows that ERβcx can bind with both ERα and ERβ (**C**). ERβcx was labeled in vitro with ^{35}S-methionine; it was then incubated with GST-fusion proteins of ERα or ERβ, adsorbed onto glutathione beads, and then eluted and separated by electrophoresis. In vitro translated products of ERβcx consisted of two proteins of 55 and 51 kDa which possibly came from different translation start sites

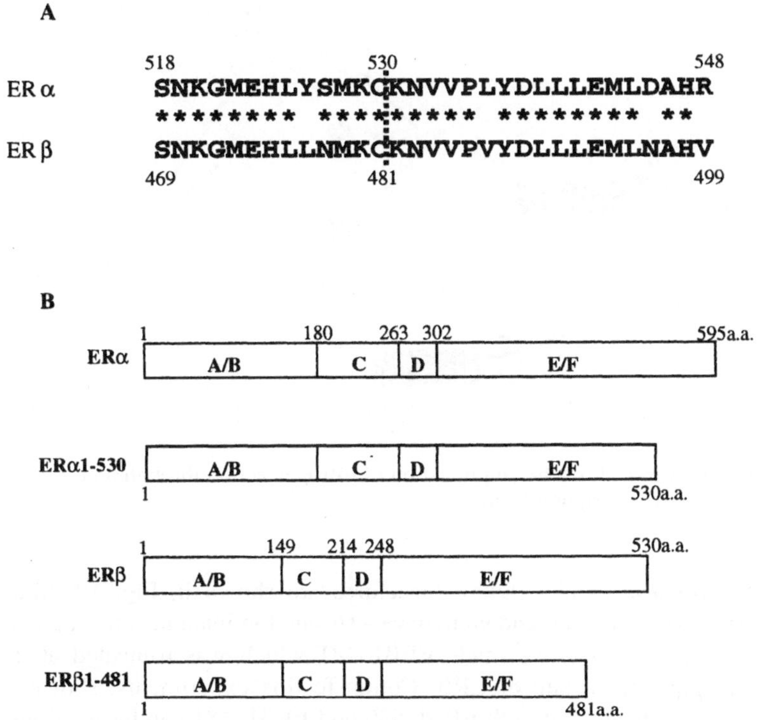

Fig. 10A,B. Structure of human ERαl–530 and ERβ1–481. Sequences around the C-terminal-truncated portion of ERαl–530 and ERβ1–481. Vertical dots show the cutting sites (**A**). Comparison of ERα, ERβ, and their C-terminal truncated molecules (**B**).

Fig. 9A shows that ERβcx interferes functionally or transcriptionally βwith ERα, whose activity is dose-dependently inhibited by the former. Interestingly, transactivation by ERβ itself does not seem to be suppressed significantly by ERβcx,(Fig. 9B). GST pull-down assay has shown that both ERα and ERβ can bind ERβcx (Fig. 9C), this confirms the finding that ERβcx may heterodimerize with ERα as well as with ERβ (Cowley et al. 1997, Pettersson et al. 1997). The cross-talk between the ERα and ERβ signaling pathways was also shown by the dominant negative activity of the C-terminal truncated molecules of both ERα and

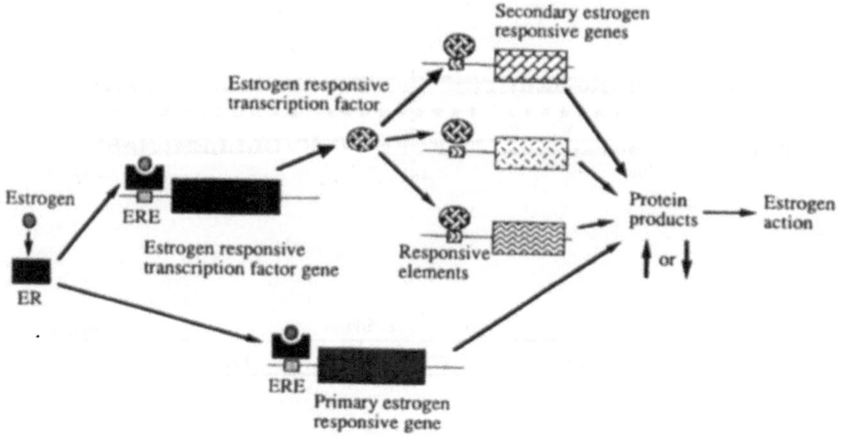

Fig. 11. A model for the amplification of estrogen action through an estrogen-responsive transcription factor

ERβ (Ogawa et al. 1998c). Structures are shown in Fig. 10. Since ERα1–530 had been known to have a strong dominant negative activity (Ince et al. 1993), we made ERβ1–481 which was truncated at the corresponding 3' site of ERβ. Cotransfection experiments, shown in Fig. 11, indicate that both ERαl–530 and ERβ1–481 can have a dominant negative action on ERα as well as on ERβ. GST pull-down assays also clearly demonstrated their molecular interaction. These findings are important because they suggest the possibility of making a transgenic animal with one of these dominant negative molecules. If it is expressed in target tissues, it should inhibit not only ERα but also ERβ, resulting in a phenotype mimicking an ERα/ERβ double-targeted animal.

In fact, we have recently produced such a transgenic rat and found it to have abnormal bone metabolism as a phenotype; this will be published elsewhere (manuscript in preparation).

Table 1. Major estrogen-responsive genes identified to contain ERE

Genes coding for:	Distribution	Targets or function
Hitherto known		
Prolactin	Pituitary (anterior lobe)	Mammary gland, milk production, secretion, CNS
Vitellogenin	Liver	Production of yolk protein
Ovalbumin	Oviduct epithelium	Egg-white protein
pS2	Cloned from MCF-7	Rather ubiquitous
Progesterone receptor	Ovarium (corpus luteum), placenta, testis	Uterus (maintenance of pregnancy), hypothalamus
Oxytocin	Pituitary (posterior lobe)	Uterus contraction, lactation, hypothalamus
Lactoferrin	Mammary gland	Iron transport in milk
Cathepsin D	Cloned from mammary tumor	Acid protease present in lysosome
Obtained by GBSC		
Efp (*E*strogen responsive *f*inger *p*rotein)	Placenta, endometrium, CNS	Growth modulation
EREG-1	Cloned from HeLa cells	A member of Mut-T family
EBAG9	Cloned from MCF-7 cells	Expressed in cervical and ovaial carcinoma
COX7RP	Ubiquitous	Cell respiration
NMDA receptor 2D	CNS, hypothalamus	Receptor for excitatory glutaminergic stimulus

3.7 Cloning and Characterization of Downstream Genes of ER

To understand the full role of estrogen in human physiology, one needs to know not only the direct effects of E_2 through ERs, but also the indirect effects governed by the estrogen-responsive genes. In fact, relatively few genes are known to be regulated directly by ER through ERE (Table 1). There are clearly not enough such genes to explain the rapid growth of the uterus at puberty, when estrogen begins to be secreted from the ovary, and to explain how the endometrial cells proliferate after the loss at menstruation, and how estrogen can prevent bones from losing calcium at menopause and even protect the brain from senile dementia (McEwen 1994, Bock and Good 1995). For this reason, we are now trying to clone and identify more downstream genes of ER (Inoue et al. 1991, Inoue et al. 1993, Orimo et al. 1995).

Various methods may be considered for this purpose, including subtractive hybridization, differential display, and so forth. We have developed a unique procedure in which a highly purified recombinant peptide of hER-DBD (DNA binding domain) is mixed with human-genomic DNA, fragmented by appropriate restriction enzymes, and passed through a nitrocellulose filter. The bound fragments are recovered, cloned in a plasmid, amplified, and again mixed with the hER-DBD. After several cycles of this procedure, the clones obtained are analyzed for the presence of ERE. It is interesting that most recovered DNA fragments thus cloned and were between 200 to 2000 bp long contained at least one ERE. We could use these probes for cloning new ERE-containing genes, which are most likely to be estrogen-responsive genes. With this procedure (Inoue et al. 1999), which is called genomic binding site cloning (GBSC), we have obtained a number of estrogen-responsive genes. These include the estrogen-responsive finger protein, Efp, which has a RING finger as well as a pair of Zn fingers (Inoue et al. 1993, Orimo et al. 1995), EREG-1 which is a human homolog of the *E. coli* Mut-T protein which probably prevents mismatch mutations by acting as dUTP phosphatase (Hiroi et al., unpublished), a cytochrome c oxidase subunit VIIa subunit-like protein (COX7RP), which may serve as an important component of mitochondrial electron transfer, and a protein named EBAG9 (Watanabe et al. 1998), which has recently been found to be expressed in human cervix and ovarial carcinoma and is possibly related to the malignancy (Sonoda et al. 1996). We have already made a knockout mouse for Efp, and found that the female has a hypoplastic uterus; a more detailed study of the phenotype is under way (manuscript in preparation). If Efp is indeed a transcription regulator, as suggested by its structure, there must be amplification of estrogen action by a cascade through ER (Fig. 11, Inoue et al. 1993).

Interestingly enough, with this method we have discovered that the gene for a glutamate receptor, NMDA receptor type 2D (NR2D), has a number of ERE half-sites at the 3' nontranslated region of mRNA; this gene is actually responsive to E_2 in the hypothalamus when E_2 is administered to ovariectomized rats (Watanabe et al. 1999). This suggests that, in some neural cells, NR2D expression could be controlled by estrogen through the ER, and that at least some part of the ER found by in situ hybridization in the CNS, especially in various nuclei in the hypothalamus (Simerly et al. 1990), may activate this type of neural receptor genes. Thus, estrogen action on the human physiology appears to be amplified not only when cognate enzymes and transcription factors are activated or increased (Inoue et al. 1993), but also when the neural receptor levels are elevated, to cause increased sensitivity, or activity, of the nerve functions.

3.8 Conclusion

I have shown here some of the recent advances in steroid research, focusing on the new developments in the study of estrogen.

Two major breakthroughs have been made in this field. The first is the identification of several coactivators (or corepressors), mediating the transactivation of transcription between the preinitiation complex at the promoter and the sequence-specific activators bound to the enhancer (e.g., ERE region). This aspect will continue to develop and will be combined with the mechanisms of transcription regulation of other genes. More cofactors must be found and their mechanisms will be the target of immediate investigation. The findings in this area will become a direct contribution to the knowledge of transcription in general.

The second is the discovery of ERβ, which apparently functions independently of ERα, which has been the only mediator of estrogen action. ERβ, which is coded by a gene distinct from that coding for

ERα, is expressed in a different spectrum of tissues and is regulated differently, although it can heterodimerize with ERα. Although the specificity of ERβ action is still unknown, it must, at least in part, cooperate with ERα to develop the female phenotype, and have gonadal functions as well, since it is expressed abundantly in both testis and ovary. The demonstration of the presence of a splice isoform ERβcx has brought another player on this stage. Just how ERβcx could have a dominant negative effect on the activation by ERα but not by ERβ remains to be determined.

There is a third important but not yet fully explored direction of estrogen research. It is the exploration of downstream genes of ER and the analysis of their role with regard to the expression of estrogen function. In the last part of this chapter I have introduced some ongoing studies in our laboratory. These investigations suggest that a large number of genes are possibly activated directly or indirectly by estrogen, in different ways, which involve transcription cascades and neural control. All these aspects will be the targets of future studies of estrogen action on humans, as well as studies on nuclear receptors in general.

References

Bock GR, Good JA (1995) Non-reproductive actions of sex steroids. Ciba Foundation Symposium 191, Wiley & Sons, Chichester

Chakravarti D, LaMorte VJ, Nelson MC, Nakajima T, Schulman IG, Juguilon H, Montminy M, Evans RM (1996) Role of CBP/P300 in nuclear receptor signaling. Nature 383:99–103

Cowley SM, Hoare S, Mosselman S, Parker MG (1997) Estrogen receptors α and β form Heterodimers on DNA. J Biol Chem 272:19858–19862

Evans RM (1988) Steroid and thyroid hormone receptors as transcriptional regulators of development and physiology. Science 240:889–895

Green S, Chambon P (1988) Nuclear receptors enhance our understanding of transcription regulation. Trends Genet 4:309

Heery DM, Kalkhoven E, Hoare S, Parker MG (1997) A signature motif in transcriptional coactivators mediates binding to nuclear receptors. Nature 387:733–736

Hiroi H et al (to be published)

Ince BA, Zhuang Y, Wrenn CK, Shapiro DJ, Katzenellenbogen BS (1993) Powerful dominant negative mutants of the human estrogen receptor. J Biol Chem 268:14026–14032

Inoue S, Kondo S, Hashimoto M, Kondo T, Muramatsu M (1991) Isolation of estrogen receptor-binding in human genomic DNA. Nucleic Acids Res 19:4091–4096

Inoue S, Orimo A, Hosoi T, Kondo S, Toyoshima H, Kondo T, Ikegami A, Ouchi Y, Orimo H, Muramatsu M (1993) Genomic binding-site cloning reveals an estrogen responsive gene that encodes a RING finger protein. Proc Natl Acad Sci USA 90:11117–11121

Inoue S, Kondo S, Muramatsu M (1999) Identification of target genes for a transcription factor by genomic binding-site cloning. In: Latchman DS (eds) Transcription factors, 2nd edn. Oxford Univ. Press, New York

Jensen EV, DeSombre ER (1973) Estrogen–receptor interaction. Science 182:126–134

Kamei Y, Xu L, Heinzel T, Torchia J, Kurokawa R, Gloss B, Lin SC, Heyman RA, Rose DW, Glass CK, Rosenfeld MG (1996) A CBP integrator complex mediates transcriptional activation and AP-1 inhibition by nuclear receptors. Cell 85:403–414

Kuiper GG, Enmark E, Pelto-Huikko M, Nilsson S, Gustafsson JA (1996) Cloning of a novel estrogen receptor expressed in rat prostate and ovary. Proc Nat Acad Sci USA 93:5925–5930

Le Douvarin B, Nielsen AL, Garnier JM, Ichinose H, Jeanmougin F, Losson R, Chambon P (1996) A possible involvement of TIF1 alpha and TIF1 beta in the epigenetic control of transcription by nuclear receptors. EMBO J 15:6701–6715

Lubahn DB, Moyer JS, Golding TS, Couse JF, Korach KS, Smithies O (1993) Alteration of reproductive function but not prenatal sexual development after insertional disruption of the mouse estrogen receptor gene. Proc Nat Acad Sci USA 90:11162–11166

Mangelsdorf DJ, Thummel C, Beato M, Herrlich P, Schutz G, Umesono K, Blumberg B, Kastner P, Mark M, Chambon P, et al (1995) The nuclear receptor superfamily: the second decade. Cell 83:835–839

McEwen BS (1994) Ovarian steroids have diverse effects on brain structure and function. In: Berg G, Hamma M (eds) The modern management of the menopause. Parthenon, Park Ridge, pp 269–278

Mosselman S, Polman J, Dijkema R (1996) ERβ: identification and characterization of a novel estrogen receptor. FEBS Lett 392:49–53

Ogawa S, Lubahn DB, Korach KS, Pfaff DW (1997) Behavioral effects of estrogen gene disruption in male mice. Proc Natl Acad Sci USA 94:1476–1481

Ogawa S, Inoue S, Watanabe T, Hiroi H, Orimo A, Hosoi T, Ouchi Y, Muramatsu M (1998a) The complete primary structure of human estrogen receptor β (hERβ) and its heterodimerization with ERα in vivo and in vitro. Biochem Biophys Res Commun 243:122–126

Ogawa S, Inoue S, Watanabe T, Orimo A, Hosoi T, Ouchi Y, Muramatsu M (1998b) Molecular cloning and characterization of human estrogen receptor βcx: a potential inhibitor of estrogen action in human. Nucleic Acids Res 26:3505–3512

Ogawa S, Inoue S, Orimo A, Hosoi T, Ouchi Y, Muramatsu M (1998c) Cross-inhibition of both estrogen receptor α and β pathways by each dominant negative mutant. FEBS Lett 423:129–132

Orimo A, Inoue S, Ikeda K, Noji S, Muramatsu M (1995) Molecular cloning, structure and expression of mouse estrogen-responsive finger protein Efp. J Biol Chem 270:24406–24413

Pettersson K, Grandien K, Kuiper GG, Gustafsson JA (1997) Mouse estrogen receptor β forms estrogen response, element-binding heterodimers with estrogen receptor α. Mol Endocrinol 11:1486–1496

Puigserver P, Wu Z, Park CW, Graves R, Wright M, Spiegelman BM (1998) A cold-inducible coactivator of nuclear receptors linked to adaptive thermogenesis. Cell 92:829–839

Rissman EF, Early AH, Taylor JA, Korach KS, Lubahn DB (1997) Estrogen receptors are essential for female sexual reactivity. Endocrinol 138:507–510

Simerly RB, Chang C, Muramatsu M, Swanson LW (1990) Distribution of androgen and estrogen receptor mRNA-containing cells in the rat brain.An in situ hybridization study. J Comp Neurol 294:76–95

Sonoda K, Nakashima M, Kaku T, Kamura T, Nakano H, Watanabe T (1996) A novel tumor-associated antigen expressed in human uterine and ovarian carcinomas. Cancer 77:1501–1509

Tremblay GB, Tremblay A, Copeland NG, Gilbert DJ, Jenkins NA, Labrie F, Giguere V (1997) Cloning, chromosomal localization and functional analysis of the murine estrogen receptor β. Mol Endocrinol 11:353–365

Voegel JJ, Heine MJ, Tini M, Vivat V, Chambon P, Gronemeyer H (1998) The coactivator TIF2 contains three nuclear receptor-binding motifs and mediates transactivation through CBP binding-dependent and independent pathways. EMBO J 17:507–519

Watanabe T, Inoue S, Ogawa S, Ishii Y, Hiroi H, Ikeda K, Orimo A, Muramatsu M (1997) Agonistic effect of tamoxifen is dependent on cell type, ERE-promoter context and estrogen receptor subtype: functional difference between estrogen receptors α and β. Biochem Biophys Res Commun 236:140–145

Watanabe T, Inoue S, Hiroi H, Orimo A, Kawashima H, Muramatsu M (1998) Isolation of estrogen responsive genes with a CpG island library. Mol Cell Biol 18:442–440

Watanabe T, Inoue S, Hiroi H, Orimo A, Muramatsu M (1999) NMDA receptor type 2D gene as target of estrogen receptor in the brain. Mol Brain Res 63:375–379

4 Global Effects of Estrogens in the Adult Central Nervous System

D.W. Pfaff

Certain effects of hormones in the central nervous system have been analyzed to a satisfying level with regard to mechanism and detail. The best understood example is that of the effects of the ovarian hormones estradiol and progesterone on female reproductive behaviors; this has been worked out in detail. The neural circuitry for the typically female primary reproductive behavior, lordosis, has been determined and the participation of estrogen-facilitated gene transcription has been documented, and has been reviewed (Pfaff 1999a). Relying heavily on this reference, we here expand the universe of the discourse on this, to consider effects of estrogens that have a much broader functional import, variety of neural sites of action, and range of mechanisms employed.

For female reproductive behavior itself, the gene for the classical estrogen receptor, ERα, was proven to be necessary for the normal performance of lordosis (Ogawa et al. 1996). The opportunities and difficulties involved in charting gene/behavior relations have also been reviewed (Pfaff 1999b; Ogawa and Pfaff, 1999). If a broader range of possibilities, beyond the classical estrogen receptor, is considered, it can be envisioned that ERβ will play an important role in brain function, as will rapid membrane effects (Moss et al. 1997).

Among the foreseeable global effects of hormones, which would be of importance for human health? One subject which clearly needs elucidation is the growth-related effect of estrogens in the brain. A second subject has to do with estrogen and mood. These two topics are discussed in this chapter, and a neural/molecular test system of potential use for medicinal chemistry related to these topics is introduced.

4.1 Neuronal Growth

The first evidence that growth-related processes in a nerve cell could be affected by estradiol came from the striking alterations in the nucleolus, the nucleus, and cell bodies in hypothalamic neurons after estrogen treatment. Biosynthetic and structural effects of steroid hormones outside the brain were plentiful and prominent, consequently stimulating interest in corresponding neurobiological investigations of estrogenic effects. Brain tissue from ovariectomized female rats either given estradiol for two weeks or given vehicle control treatment was perfused, to properly fix cells in the hypothalamus for electron microscopy. From the first publication (Cohen and Pfaff 1981), and extending for several years, this line of research turned out to be a rich source of new insights related to trophic actions of steroid hormones in the brain.

These first morphological, structural effects of estrogen were revealed to be a massive elaboration of the rough endoplasmic reticulum in hypothalamic nerve cells, in the ventromedial region where neurons have estrogen receptors, as well as a large increase in the number of prosecretory granules in the vicinity of the Golgi apparatus. All of these findings pointed to an increase in the biosynthetic activity of these hypothalamic nerve cells with estrogen receptors. Subsequently, Chung et al. showed alterations on the surface of the nucleolus, where ribosomal RNA is synthesized (Chung et al. 1984; Cohen et al. 1984). This type of work was reviewed (Cohen and Pfaff 1992). Estrogen apparently increases the appearance of ribosomal RNA's gene on the surface of the nucleolus, where rRNA would be synthesized. We followed up the possibility of the ribosomal RNA alterations, and demonstrated not only the structural effects of estrogens acting in the nucleus (Jones et al. 1985), but also showed that there are increased molecular hybridization products, by using radioactively labeled probes for in situ hybridization, which would measure either newly synthesized (external transcribed spacer in the nascent transcript) or fully mature ribosomal RNA (Jones et al. 1986; Jones et al. 1990). Overall, the rapid molecular effects of estrogen on the generation of ribosomal RNA preceded the eventual structural alterations in hypothalamic neurons.

A natural result of this increased capacity for synthetic activity in the "power plant" of the neuron is to increase the number of synapses. Chung et al. (1988) were able to see an increase in the number of synapses from the hypothalamus upon estrogen treatment, a finding reminiscent of the report by Carrer and Aoki (1982). This type of work was extended by Frankfurt et al., who found an increased number of dendritic spines on hypothalamic neurons (Frankfurt 1994; Gould et al. 1990).

This field of morphological work has demonstrated that estrogen treatment can enhance the potential of hormone-sensitive hypothalamic neurons to produce proteins and to grow, and that estrogen also affects the structural basis for inputs and outputs of hypothalamic neurons to and from other cells. The full set of mechanisms for these changes has not yet been charted.

A surprising manifestation of hypothalamic neuronal activity is that it helps to maintain the health of certain neurons in the midbrain central gray. Destruction of ventromedial hypothalamic cells can cause transsynaptic degeneration in the midbrain (Chung et al. 1990).

To examine possible implications for sex steroid effects on memory, it was natural to examine the hippocampus. Estrogen can increase the number of dendritic spines in hippocampal pyramidal cells (Gould, Woolley et al. 1990; Woolley et al. 1990; Woolley and McEwen 1992). Other recent evidence, as well, shows that trophic actions of estrogens can be applied outside the hypothalamus. In female prairie voles, cell division in the ventricular and subventricular zones was facilitated by estrogen treatment, as shown by an increased number of BrdU-labeled cells (Smith et al. 1997).

A growth-related effect of estradiol was also reported to occur in a part of the brain never before associated with sex hormone action. In the cat brain, the nucleus retroambiguus is located in the caudal medulla and sends axons into the spinal cord, where they terminate amongst pelvic floor and other axial motoneurons. It was surprising to see that the density of axonal terminations among their target motoneuronal cell groups, from this descending system, was much greater in the presence of high levels of estrogen (Vanderhorst and Holstege 1997). As a neuroanatomical control for specificity of the hormone effect, no differences were found in the rubrospinal pathway. Labeled growth cones were seen in the descending retroambiguus pathways of estrous, but not nonestrous cats. This paper was the first demonstration of trophic actions of estrogen in a brainstem system descending to the spinal cord.

4.2 Mechanisms

The primary cellular mechanisms underlying the protective effects of estrogens on cognitive function are likely to be the first ones discovered: the growth-related actions of estrogen on the "power plant" of neurons, as reviewed above. Theoretically, if all the protein-synthetic machinery of a neuron, including both the nucleus and the cytoplasm is enhanced (Chung, Cohen et al. 1984; Cohen and Pfaff 1981; Jones, Harrington et al. 1990; Jones, Pfaff et al. 1985), it would follow that dendrites and axons can grow, synapses can be formed, and the cognitive functions that depend on those synapses can flourish.

Connections between steroid hormones and growth factors have been reported. Strikingly, the results of Dominique Toran-Allerand (Toran-Allerand et al. 1983) and Robert Gibbs (Gibbs et al. 1994) in the CNS on NGF and its receptors are consistent with those of Kenneth Korach (Washburn et al. 1991), who in 1991 pointed out the effects of EGF on estrogen actions in the uterus. Toran-Allerand, who earlier had reported sprouting of cell processes in response to hormonal treatment, now has shown that there is a degree of cooperation between estrogens and nerve growth factor (NGF), each possibly amplifying the other's effects on nerve cell growth. Recently, it has been reported that estradiol can stimulate tyrosine phosphorylation of an insulin-like growth factor receptor (Richards et al. 1996). Conversely, neurotrophic factors can stimulate the expression of a sex hormone receptor in neurons (Al-Shamma and Arnold 1997).

An entirely new possibility has recently come up for helping to explain estrogen-stimulated neuronal growth. This derives from the estrogen sensitivity of vascular endothelial growth factor (V-EGF) (Hyder et al. 1996; McLaren et al. 1996; Shifren et al. 1996). The induction of V-EGF by estrogen is ER-mediated, as a pure antiestrogen will block it (Chiapetta et al. 1997). Better vascularity could be part of the way in which estrogen can help protect against strokes (Paganini-Hill 1995), β-amyloid toxicity (Behl et al. 1996; Goodman et al. 1996; Green et al. 1996), oxidative damage (Dluzen et al. 1996), and the consequences of penetrating brain injury (Garcia-Estrada et al. 1993). So, estrogenic effects on the cerebral vasculature could be important for hormonal facilitation of neuronal growth and repair.

One component of the growth cone in actively growing axons is the phosphoprotein growth-associated protein GAP-43. Lustig et al. (1991) found 48–74% increases in the mRNA for GAP-43 in the ventromedial hypothalamus within two hours of estrogen treatment, an effect likely to represent a stimulation of GAP-43 gene transcription by estrogen. More recently, Singer and Dorsa (Singer et al. 1996) also noted estrogenic stimulation of GAP-43 mRNA in the preoptic area and found that an age-related decline could be reversed by estrogen treatment. Clearly, estrogenic actions such as this could contribute to the cognitive effects of the hormone.

Revealing still another type of mechanism, Luine (Luine 1985; Luine 1997) and her colleagues have shown that estrogen administration to ovariectomized female rats can increase the levels of the enzyme that produces the transmitter acetylcholine (ACh). This is a potentially important finding, since Alzheimer's disease patients suffer a massive loss of ACh neurons in their basal forebrains. The complex of changes in cholinergic neuronal molecular mechanisms and in the mRNAs for nerve growth factor receptors may possibly contribute to the behavioral results of interest (Gibbs, Wu et al. 1994; McMillan et al. 1996). Still another potential mechanism concerns the receptor for the excitatory transmitter-like substance, NMDA (*N*-methyl-D-aspartate). For example, Morrison and Weiland found that female rats treated with estrogen had 30% more NMDA binding in certain hippocampal neurons.

Finally, a growth factor very recently recognized as having possible importance for hormone effects on neuronal growth is brain-derived neurotrophic factor (BDNF). The most telling new experiment was launched following the demonstrations that estradiol can induce an increase in dendritic spine density on hippocampal neurons in vivo and in vitro (Murphy and Segal 1996; Woolley and McEwen 1992). BDNF, involved in the regulation of GABA interneurons, is down-regulated by estradiol in hippocampal neurons (Murphy et al. 1998). Thus, neuronal inhibition is decreased, and excitation is increased among hippocampal pyramidal neurons. As a result, there is a twofold increase in dendritic spine density. Exogenous BDNF blocks estrogenic growth-related effects in hippocampus, while antisense DNA administration against BDNF mRNA has the opposite effect. BDNF is thus drawn into the mechanisms by which estrogens act in the hippocampus (Murphy, Cole et al. 1998).

In summary, morphological, neurochemical, and molecular findings have begun to provide explanations of cognitive enhancement due to estrogen treatment.

4.3 Estrogens and Mood

Estrogens can significantly improve mood, by helping to combat depression, as has been documented during hormone replacement therapy (Aylward 1976; Daly et al. 1993; Limouzin-Lamothe et al. 1994; Michael et al. 1970; Rubinow and Schmidt 1996; Stahl 1996). This phenomenon is not necessarily limited by culture, as Chinese women with postpartum depression were found to have decreased levels of estradiol in the blood. (Lu et al. 1996). The power of estrogen in fighting depression is robust; it can reduce postpartum affective disorder in women with a history of puerperal major depression (Sichel et al. 1995). Admittedly, the effects of estrogen on mood can be context-dependent and may vary according to the individual. For example, in ten women with severe premenstrual mood changes, estrogen administration was, unfortunately, associated with a recurrence of symptoms (Schmidt et al. 1998).

Before further progress can be made in this field, there is a pressing need to design animal models for hormone effects on mood, so that new, potentially therapeutic compounds can be tested rapidly. Toward this end, we have uncovered a striking effect of estrogens on a behavioral test proven sensitive to the effects of antidepressants. The Porsolt Forced Swim Test (FST) consists of two sessions during which rats are placed in water-filled cylinders; their behavior is then put in one of the following three categories: swimming, struggling, or immobility. Animals who show the presumed analog of "depression" quickly give up, and collapse into passive immobility. Animals given antidepressants will spend significantly longer times swimming. On test day 1, our estradiol-treated ovariectomized female rats spent significantly less time struggling, and virtually no time in immobility – in contrast, they spent significantly more time with normal swimming than the vehicle-treated ovariectomized rats did (Rachman et al. 1998). On test day 2, the differences between estrogen-treated and control animals were even more pronounced. Moreover, induction of c-fos immunoreactivity in neurons, as stimulated by the fear and effort of the test, was significantly reduced in the estrogen-treated groups in broad areas of the forebrain related to sensory, contextual, and integrative processing. These results (Rachman et al. 1998) are consistent with the estrogen effects on mood as reported in women, and may offer a convenient test system for future use.

4.4 Estrogen-Stimulated Transcription in a Neuroendocrine Cell Line

For thorough analysis of the mechanisms of potentially therapeutic estrogen-like compounds, a sensitive, reliable transcription system in neurons is required. After all, the neural parenchyma presents the molecular biologist with very small amounts of starting material and the problem of cellular heterogeneity. The immortalized neuroendocrine cell line GT-1–7, produced from GnRH neurons by targeted oncogenesis, offers a potential solution to this problem. These nerve cells, lacking their own estrogen-binding capacity, can be transfected with the ER of choice and be used for quantitative studies of estrogenic effects on transcription (Attardi et al. 1998). In this new work, GT-1–7 cells were transfected with ERα, or with ERα mutated in its DNA-binding domain, or with no ER. As a reporter gene, the luciferase gene was placed under the control of a triple ERE. Following a 20-hour incubation period with various estrogenic or control compounds, cells were harvested and luciferase activity was measured. Under these circumstances, an orderly dose–response curve for estradiol 17β was obtained. The transcriptional activation through the ERE was shown to depend not only on the presence of ER, but also on its normal DNA-binding capacity. The surprising result was the poor correlation between ER-binding affinity and the degree of transactivation, across a range of 12 estrogenic compounds. For example, the compound 11β-acetoxyestradiol had a greater transcriptional effect than estradiol, but much worse relative binding affinity. In the opposite direction, dienestrol had only one half the transcriptional effect of estradiol but had more than twice the ER-binding affinity (Attardi et al. 1998).

In these neuroendocrine cells, therefore, factors other than affinity for the classical ER, clearly, are helping to determine the molecular effects of various estrogens. This test system might be quite revealing as new approaches to hormone response therapies involving the brain are sought.

References

Al-Shamma H, Arnold A (1997) Brain-derived neurotrophic factor regulates expression of androgen receptors in perineal motoneurons. Proc Natl Acad Sci USA 94:1521–1526

Attardi B, Pfaff D, Hendry L (1998) Transcriptional consequences of estrogens in GT-1 neurons are not correlated with classical estrogen receptor affinities. Society for Neuroscience, Abstract

Aylward M (1976) Estrogens, plasma tryptophan levels in perimenopausal patients. In: Campbell S (ed) The management of the menopause and post-menopausal years. Univ. Park Press, pp 135–147

Behl C, Widmann M, Trappand T, Holsboer F (1996) 17β estradiol protects neurons from oxidative stress-induced cell death in vitro. Biochem Biophys Res Commun 216:473–482

Carrer H, Aoki A (1982) Ultrastructural changes in the hypothalamic ventromedial nucleus of ovariectomized rats after estrogen treatment. Brain Res 240:221–233

Chiapetta C, Murthy L, Stancel G, Hyder S (1997) The pure antiestrogen ICI 182, 780 differentially inhibits induction of two estrogen regulated genes, c-fos and vascular endothelial growth factor (Abstract P3-416). In: Endo '97: The Endocrine Society, p 540

Chung S, Cohen R, Pfaff D (1984) Ultrastructure and enzyme digestion of nucleoli and associated structures in hypothalamic nerve cells viewed in resinless sections. Biol Cellulaire 51:23–34
Chung S, Pfaff D, Cohen R (1988) Estrogen-inducted alterations in synaptic morphology in the midbrain central gray. Exper Brain Res 69:522–530
Chung S, Pfaff D, Cohen R (1990) Projections of ventromedial hypothalamic neurons to the midbrain central gray: An ultrastructural study. Neuroscience 38:395–407
Cohen R, Pfaff D (1981) Ultrastructure of neurons in the ventromedial nucleus of the hypothalamus in ovariectomized rats with or without estrogen treatment. Cell and Tissue Res 217:451–470
Cohen R, Pfaff D (1992) Ventromedial hypothalamic neurons in the mediation of long-lasting effects of estrogen on lordosis behavior. Prog Neurobiol 38:423–453
Cohen R, Chung S, Pfaff D (1984) Alteration by estrogen of the nucleoli in nerve cells of the rat hypothalamus. Cell & Tissue Res 235:485–489
Daly E, Gray A, Barlow D, Mcpherson K, Roche M, Vessey M (1993) Measuring the impact of menopausal symptoms on quality of life. Br Med J 307:836–840
Dluzen D, McDermott J, Liu B (1996) Estrogen as a neuroprotectant against MPTP-induced neurotoxicity in C57/B1 mice. Neurotoxicol Teratol 18:603–606
Frankfurt M (1994) Gonadal steroids and neuronal plasticity: studies in the adult rat hypothalamus. Ann NY Acad Sci 743:45–60
Garcia-Estrada J, Del Rio J, Luquin S, Soriano E, Garcia-Segura M (1993) Gonadal hormones down-regulate reactive gliosis and astrocyte proliferation after a penetrating brain injury. Brain Res 628:271–278
Gibbs R, Wu D-H, Hersh L, Pfaff D (1994) Effects of estrogen replacement on relative levels of choline acetyltransferase, trkA, and nerve growth factor messenger RNAs in the basal forebrain and hippocampal formation in adult rats. Exp Neurol 129:70–80
Goodman Y, Bruce A, Cheng B, Mattson M (1996) Estrogens attenuate and corticosterone exacerbates excitotoxicity, oxidative injury, and amyloid beta-peptide toxicity in hippocampal neurons. J Neurochem 66:1836–1844
Gould E, Woolley C, Frankfurt M, McEwen B (1990) Gonadal steroids regulate dendritic spine density in hippocampal pyramidal cells in adulthood. J Neurosci 10:1286–1291
Green P, Gridley K, Simpkins J (1996) Estradiol protects against β-amyloid (25–35)-induced toxicity in SK-N-SH human neuroblastoma cells. Neurosci Lett 218:165–168

Hyder S, Stancel G, Chiappetta C, Murthy L, Boettger L, Makela S (1996) Uterine expression of vascular endothelial growth factor is increased by estradiol tamoxifen. Cancer Res 56:3954–3960

Jones K, Pfaff D, McEwen B (1985) Early estrogen-induced nuclear changes in rat hypothalamic ventromedial neurons: an ultrastructural and morphometric analysis. J Comp Neurol 239:255–266

Jones K, Chikaraishi D, Harrington C, McEwen B, Pfaff D (1986) In situ hybridization detection of estradiol-induced changes in ribosomal RNA levels in rat brain. Mol Brain Res 1:145–152

Jones K, Harrington C, Chikaraishi D, Pfaff D (1990) Steroid hormone regulation of ribosomal RNA in rat hypothalamus: Early detection using in situ hybridization and precursor-product ribosomal DNA probes. J Neurosci 10:1513–1521

Limouzin-Lamothe M, Mairon N, LeGal J, LeGal M (1994) Quality of life after the menopause: influence of hormonal replacement therapy. Am J Obstet Gynecol 170:618–624

Lu R, Ko H, Yao B, Chang F, Yeh T, Huang K (1996) Lower level of estradiol in postpartum depression among Chinese women. Biol Psychiatry (Abstract 504) 39:648

Luine V (1985) Estradiol increases choline acetyltransferase activity in specific basal forebrain nuclei and projection areas of female rats. Exp Neurol 89:484–490

Luine V (1997) Estrogenic enhancements of and sex differences in memory in rats. Soc Behav Neuroendo 44

Lustig R, Sudol M, Pfaff D, Federoff H (1991) Estrogenic regulation and sex dimorphism of growth-associated protein 43 kDa (GAP-43) messenger RNA in the rat. Mol Brain Res 11:125–132

McLaren J, Prentice A, Charnock-Jones D, Millican S, Muller K, Sharkey A, Smith S (1996) Vascular endothelial growth factor is produced by peritoneal fluid macrophages in endometriosis and is regulated by ovarian steroids. J Clin Invest 98:482–489

McMillan P, Singer C, Dorsa D (1996) The effects of ovariectomy and estrogen replacement on trkA and choline acetyltransferase mRNA expression in the basal forebrain of the adult female sprague-Dawley rat. J Neurosci 16:1860–1865

Michael C, Kantor H, Shore H (1970) Further psychometric evaluation of older women: the effect of estrogen administration. J Gerontol 25:337–341

Moss R, Gu Q, Wong M (1997) Estrogen: nontranscriptional signaling pathway. Rec Prog Horm Res 52:1–37

Murphy D, Segal M (1996) Regulation of dendritic spine density in cultured rat hippocampal neurons by steroid hormones. J Neurosci 16:4059–4068

Murphy D, Cole N, Segal M (1998) Brain-derived neurotrophic factor mediates estradiol-induced dendritic spine formation in hippocampal neurons. Proc Natl Acad Sci USA 95:11412–11417

Ogawa S, Pfaff D (1999) Genes participating in the control of reproductive behaviors. In: Pfaff DW, Berrettini W, Joh T, Maxson S (eds) Genetic influences on neural and behavioral function. CRC Press, Boca Raton

Ogawa S, Taylor J, Lubahn DB, Korach KS, Pfaff DW (1996) Reversal of sex roles in genetic female mice by disruption of estrogen receptor gene. Neuroendocrinology 64:467–470

Paganini-Hill A (1995) Estrogen replacement therapy and stroke. Prog Cardiovasc Dis 38:223–242

Pfaff DW (1999a) Drive: neural and molecular mechanisms for sexual motivation. MIT Press, Cambridge

Pfaff DW (1999b) Introduction: genetic influences on the nervous system and behavior. In: Pfaff DW, Berrettini W, Joh T, Maxson S (eds) Genetic influences on neural and behavioral function. CRC Press, Boca Raton

Rachman I, Unnerstall J, Pfaff D, Cohen R (1998) Estrogen alters behavior and forebrain c-fos expression in ovariectomized rats subjected to the forced swim test. Proc Natl Acad Sci USA 95:13941-13946

Richards R, DiAugustine R, Petrusz P, Clark G, Sebastian J (1996) Estradiol stimulates tyrosine phosphorylation of the insulin-like growth factor-1 receptor and insulin receptor substrate-1 in the uterus. Proc Natl Acad Sci USA 93:12002–12007

Rubinow D, Schmidt P (1996) Reproductive hormones and mood in women. Biol Psychiatry (Abstract 386) 39:613

Schmidt P, Nieman L, Danaceau M, Adams L, Rubinow D (1998) Differential behavioral effects of gonadal steroids in women with and in those without premenstrual syndrome. N Engl J Med 338:209–216

Shifren J, Tseng F, Zaloudek C, Ryan I, Meng Y, Ferrara N, Jaffe R, Taylor R (1996) Ovarian steroid regulation of vascular endothelial growth factor in the human endometrium: implications for angiogenesis during the menstrual cycle and in the pathogenesis of endometriosis. J Clin Endocrinol Metab 81:3112–3118

Sichel D, Cohen L, Robertson L, Ruttenberg A, Rosenbaum J (1995) Prophylactic estrogen in recurrent postpartum affective disorder. Biol Psychiatry 38:814–818

Singer C, Rogers K, Strickland T, Dorsa D (1996) Estrogen protects primary cortical neurons from glutamate toxicity. Neurosci Lett 212:13–16

Smith M, Zuoxin W, Luskin M, Insel T (1997) Estrus induction associated with neurogenesis in the brain of female prairie voles. Soc Behav Neurosci 110:174

Stahl S (1996) Reproductive hormones as adjuncts to psychotropic medications. Biol Psychiatry (Abstract 388) 39:613

Toran-Allerand C, Hashimoto K, Greenough W, Saltarelli M (1983) Sex steroids and the development of the newborn mouse hypothalamus and preoptic area in vitro: III. Effects of estrogen on dendritic differentiation. Brain Res 283:97–101

Vanderhorst V, Holstege G (1997) Estrogen induces axonal outgrowth in the nucleus retroambiguus-lumbosacral motoneuronal pathway in the adult female cat. J Neurosci 17:1122–1136

Washburn T, Hocutt A, Brautigan D, Korach K (1991) Uterine estrogen receptor in vivo: phosphorylation of nuclear specific forms on serine residues. Mol Endo 5:235–242

Woolley C, McEwen B (1992) Estradiol mediates fluctuation in hippocampal synapse density during the estrous cycle in the adult rat. J Neurosci 12:2549–2554

Woolley C, Gould E, Frankfurt M, McEwen B (1990) Naturally occurring fluctuation in dendritic spine density on adult hippocampal pyramidal neurons. J Neurosci 10:4035–4039

5 HRT and Brain Function – Clinical Aspects

T. Ohkura, K. Isse, K. Tanaka, K. Akazawa, M. Hamamoto, and N. Iwasaki

It is well established that estrogen replacement therapy (ERT) or hormone replacement therapy (HRT) is effective not only for improving menopausal symptoms, but also for preventing osteoporosis and cardiovascular disease in postmenopausal women (Belchetz 1994). Furthermore, estrogen does influence brain function, but the effects of estrogen on brain function such as memory, cognitive function, and brain blood flow have not been well established in postmenopausal women or female patients with dementia of the Alzheimer type (DAT). Our studies will to some extent clarify the clinical aspects of HRT and brain function. Based on our data, this paper will report the effects of HRT on brain function in postmenopausal women and DAT patients. Our studies are divided into three parts: 1. Estrogen and Memory (Sect. 5.1), 2. Estrogen and Brain Blood Flow (Sect. 5.2), and 3. HRT for Dementia of the Alzheimer Type (DAT) (Sect. 5.3).

5.1 Estrogen and Memory

Acetylcholine is the most important neurotransmitter for memory and learning. Estrogen acts on cholinergic neurons in several ways in the rat brain. It increases the activity of choline acetyltransferase (ChAT) (Luine 1985), and augments the production of nerve growth factor (NGF) (Gibbs et al. 1994) and brain-derived neurotrophic factor (BDNF) receptors (Singh et al. 1995). It also influences the neuronal structures of the hippocampus (Gould et al. 1990; Woolley et al. 1997) which plays a very important role in memory and learning. Some clinical studies showed that estrogen treatment improves memory function in postmenopausal women (Phillips and Sherwin 1992; Kampen and Sherwin 1994). In a study of surgically menopausal women, Phillips and Sherwin (1992) found no postoperative decline in the performance of episodic verbal memory tasks (paired-associates learning) by estradiol-treated women, but scores declined significantly in a placebo group. If that is the case, we will be able to observe a decline in the memory function of women in the climacterium, when estrogen secretion from ovaries decreases gradually and stops permanently in the end (menopause). There are many reports on the relation between aging and memory function, based on studies where men and women are grouped together. However, there are very few publications on memory tests performed on climacteric and periclimacteric women, divided into five-year-long age groups.

The present studies were designed to investigate the relation between memory function and aging in women (study 1), and the effects of ERT/HRT on memory in both postmenopausal and amenorrheic women (study 2). Furthermore, we will report on a case of premature menopause accompanied by a memory disorder after the discontinuation of HRT (study 3). The ERT and HRT used in these studies are as follows: (1) ERT: a subject receives 1.25 mg/day of conjugated equine estrogen (CEE) for 21 days, followed by a pause of 7 days. (2) HRT: a subject receives 1.25 mg of CEE from days 1–25 and 10 mg of medroxyprogesterone acetate (MPA) from days 14–25 of estrogen treatment, followed by a pause of 5 days.

5.1.1 Study 1. Memory Function and Aging

5.1.1.1 Subjects and Methods

Two hundred and forty women of our outpatient clinic, who lived a normal, ordinary life, were divided into 8 groups: A (20–29 years, n=30), B (30–34 years, n=27), C (35–39 years, n= 31), D (40–44 years, n=32), E (45–49 years, n=25), F (50–54 years, n=34), G (55–59 years, n=32), and H (60–64 years, n=29). Women under 40 years of age in the groups A, B, and C had a normal menstrual cycle. The memory function of each group was determined with Miyake's Memory Test, which consists of ten paired easy-associates and ten paired hard-associates. The easy-associates are related paired learning words (e.g., tobacco – match) and the hard-associates are unrelated paired learning words(e.g., boy – mat). One starts with the easy-associates. A tester reads ten paired words slowly, and a subject remembers these paired words. After having finished reading ten paired words, the tester says one of the paired words and the subject has say the other. The time limit is about 30 seconds per paired words. After having finished the first trial of ten questions and answers, the tester tries the second presentation of ten paired words in the same way as the first testing, and the second trial of questions and answers should be performed. If the subject answers the ten other words correctly at the first trial, the second and third trials will be omitted. The full score or number of all correct answers is 10. To repeat the memory test for the same person, five different questionnaires of ten paired words in both easy- and hard-associates are prepared. The mean scores for both easy- and hard-associates at the third trial of presentation were compared among the eight age groups.

5.1.1.2 Results

At the third trial of easy-associates, there were no significant differences in the mean scores among the groups A, B, C, D, E, F, and G. The mean score for group H was lower than those for groups A, B, C, D, E, F, and G (p<0.01) (Table 1). There were no significant differences in the mean scores among groups A, B, and C. The mean scores for these three groups were significantly higher than those for the other remaining groups. The decline of memory function was observed for the first time in group D, and then it was observed in group F. The former corresponds to the early climacteric age group in which serum estrogen levels begin

Table 1. Mean scores for easy-associates of the 3rd trials of presentation

Age groups	A	B	C	D	E	F	G	H
Correct	9.9±0.3	9.9±0.3	10.0±0.2	9.9±0.2	10.0±0.2	9.7±0.6	9.8±0.5	9.3±1.1*

mean±SD (rounded to two decimal places). *$p<0.01$ vs. groups A–G

to decrease, and the latter corresponds to the menopausal age group in which the production of ovarian estrogen stops permanently. Memory impairment progressed gradually in postclimacteric women (Fig. 1).

Significant differences in the mean scores of all the groups were most frequent and greatest at the third trial of hard-associates. Therefore, to differentiate between memory abilities of each group, the mean scores at the 3rd trial of hard-associates were compared among the age groups.

5.1.2 Study 2. The Effects of ERT/HRT on Memory

HRT was discontinued in another 19 women aged 20–39 (mean±SD: 31.8±7.3 years) with premature menopause, surgical menopause, and severe amenorrhea (serum estradiol pg/ml), who had been receiving

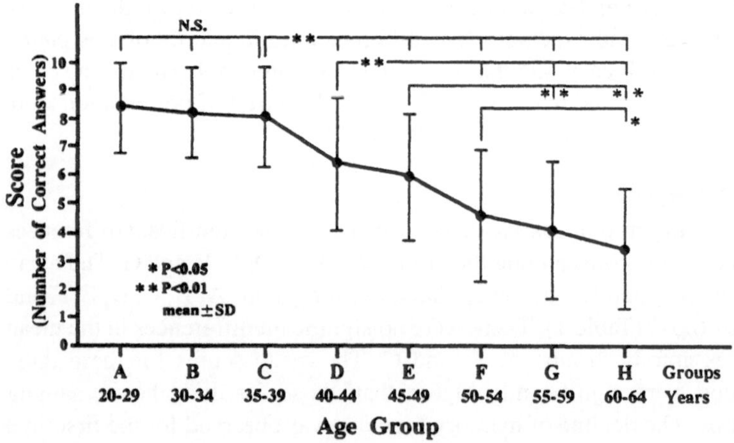

Fig 1. Mean scores for age groups. The scores are for hard-associates after three trials of presentation (*$p<0.05$, **$p<0.01$)

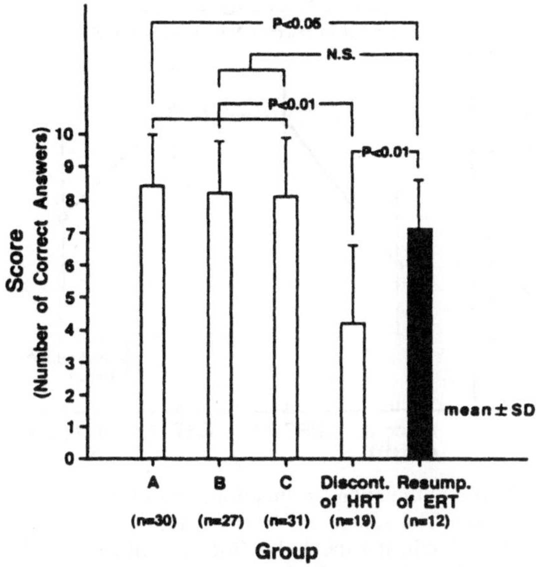

Fig 2. Effects of discontinuation and resumption of ERT/HRT on memory function. The scores are for hard-associates after three trials of presentation

HRT; 4–12 (8.3±3.0) weeks after the discontinuation of HRT, their memory function was determined with the same memory test used in study 1. A similar test was performed after ERT resumption in 12 out of the 19 women. Their mean scores were compared with those of normal age groups A, B, and C. The mean score for the HRT-discontinued group was 4.2±2.4 (SD), which was significantly lower than those of groups A, B, and C ($p<0.01$) (Fig. 2). The mean score for the ERT-resumed group was 7.1±1.5, which was lower than that of group A ($p<0.05$), but did not differ from those of groups B and C, and was significantly higher than that of the HRT-discontinued group (Fig. 2). In eight out of the twelve women whose scores for ten paired hard-associates were lower than 5, ERT and memory tests were repeatedly performed. During ERT, the mean scores for this group increased significantly ($p<0.01$), but the discontinuation of ERT again decreased the mean score (Fig. 3).

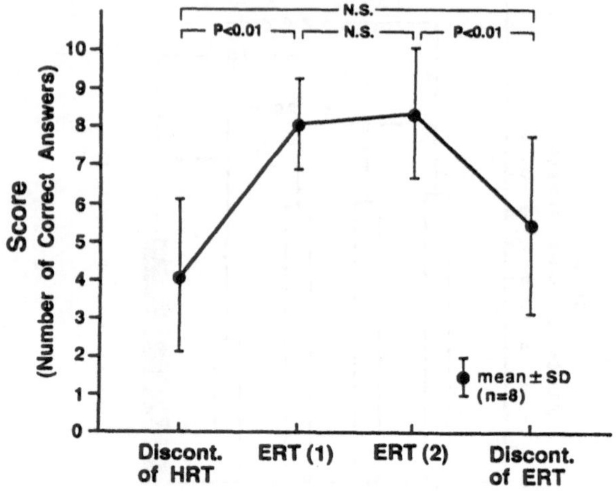

Fig 3. Effects of discontinuation, resumption, and continuation of ERT/HRT on memory function. The scores are for hard-associates after three trials of presentation. ERT (2) indicates the 2nd or 3rd cycle of ERT. Each hard-associate consisted of totally different paired words

5.1.3 Study 3. A Case of Premature Menopause Accompanied by a Memory Disorder after Discontinuation of HRT

This is a 27-year-old patient's case of premature menopause accompanied by memory disorder after the discontinuation of HRT. HRT was discontinued in order to induce ovulation; this failed. After three months of discontinuation of HRT, she complained spontaneously as follows: "Recently, my hair fell out very much, and I became forgetful. I particularly couldn't recall people's names when I was working." The memory test was performed on her two months after a five-month discontinuation of HRT. I was surprised at her low scores of correct answers, which were as low as those of an old person.

The Roman numerals in Fig. 4 indicate totally different tests of paired words. The discontinuation of HRT resulted in a decline of memory function not only in hard-associates, but also in easy-associates. The resumption of ERT increased the scores in both. The discontinuation of ERT again resulted in a decrease in the scores of hard-associates, but not in those of easy-associates this time.

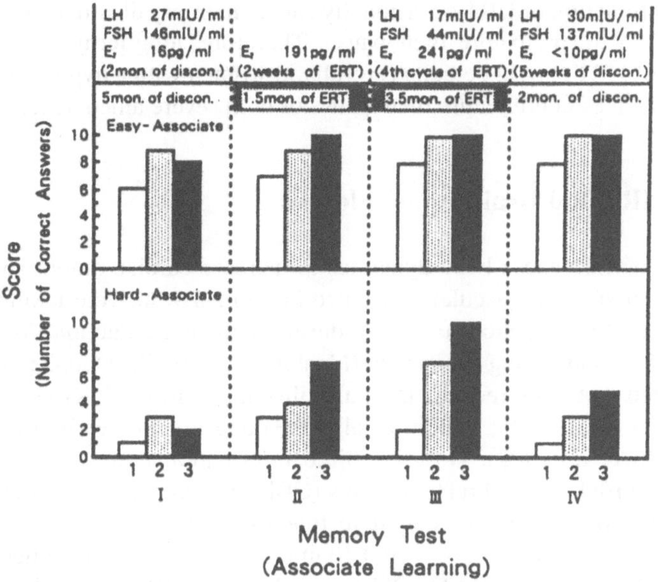

Fig 4. Effects of discontinuation, resumption, and continuation of ERT/HRT on memory function in a case of premature menopause accompanied by a memory disorder after discontinuation of HRT. The patient was 27 years old. *Roman numerals* indicate totally different tests of paired words. The date on which the serum for determining hormone levels was obtained was not always the same date on which the memory test was performed

5.1.4 Conclusion

There were no significant differences in the memory function among the three age groups before the age of 40 years (20–39 years).However, memory function declined gradually with age in climacteric and postclimacteric women. It is inferred from the results of studies 1 and 2 that a decrease in ovarian estrogen as well as aging may participate in the decline of memory function in climacteric women, and that memory function in women is closely related to serum estrogen levels. Study 2 shows that discontinuation of ERT or HRT may cause a decrease in memory function, and that the resumption of ERT improves memory function in both postmenopausal and severe amenorrheic women. The discontinuation of HRT occasionally causes memory disorders in young women with premature menopause. Therefore, long-term discontinuation of HRT should be done carefully in women with a hypoestrogenic state such as in premature menopause or with severe amenorrhea.

5.2 HRT and Brain Blood Flow

Estrogen influences hemodynamics to increase cardiac output and decrease peripheral vascular resistance in women who were taking oral contraceptives (Lehtovirta 1974), during ovulation induction (Veille et al. 1986), and during pregnancy (Mashini et al. 1987). Doppler studies revealed that ERT reduces the pulsatility index in the internal carotid (Gangar et al. 1991; Penotti et al. 1993) and middle cerebral arteries (Penotti et al. 1993). However, little information is available on the effect of ERT on cerebral blood flow (CBF). We previously reported that ERT not only improves cognitive function, but that it also increases regional cerebral blood flow (rCBF) in female patients with dementia of the Alzheimer type (DAT) (Ohkura et al. 1994a). This rCBF, however, was expressed as relative N-isopropyl-*p*-iodoamphetamine ([^{123}I]IMP) uptake (cortex/cerebellum ratio). Hence, an increase in rCBF during ERT in our previous study does not mean an increase in absolute CBF values.

We recently reported that ERT increases the whole cerebral and cerebellar blood flows (Ohkura et al. 1995a) and regional cerebral blood flow (Ohkura et al. 1996), which were measured by a quantitative method by [^{123}I]IMP and single-photon emission computed tomography (SPECT) in nine postmenopausal women (mean age±SE: 43.7±2.5 years), including young premature menopausal and surgically menopausal women. Whole cerebral and cerebellar flow (CBF and C_{bl}BF) measurements were performed four weeks after the discontinuation of HRT. After the first brain blood flow (BBF) measurements, nine subjects received 0.625 mg of CEE orally twice daily, for three continuous weeks. The second BBF measurements were performed on all nine subjects between two and three weeks after the first ones. Both the mean CBF and C_{bl}BF were significantly increased during ERT (p=0.0382). The mean percent increases were 29.5±10.2% in CBF and 29.3±10.4%

in $C_{bl}BF$. These results suggest that ERT significantly increases the whole cerebral and cerebellar blood flows in postmenopausal women.

Regional cerebral blood flow (rCBF) measurements were performed on 13 regions of each hemisphere in the above-mentioned nine postmenopausal women. The rCBF was measured in a total of 26 regions of the brain. The mean rCBF values were significantly increased in all the regions of the brain but two (left lower frontal and right upper occipital regions) during ERT ($p<0.05$). The mean percent increases (mean±SE) in the regions where a significant increase in rCBF was observed were 23.3±9.5% to 34.3±11.7%. These results suggest that ERT significantly increases rCBF in most regions of the brain in postmenopausal women.

In our new studies, we have measured brain blood flow in naturally menopausal women; young premature menopausal women were not included in these studies.

5.2.1 Study 1. Effects of Short- and Long-Term ERT on CBF and $C_{bl}BF$ in Postmenopausal Women

These studies were designed to investigate the effects of short- and long-term ERT on CBF and $C_{bl}BF$ in postmenopausal women. The CBF and $C_{bl}BF$ were measured quantitatively through the use of ^{99m}Tc-labeled ethyl cysteinate dimer (ECD) (Matsuda et al. 1995) and SPECT.

5.2.1.1 Short-Term Low-Dose and High-Dose ERT

The CBF and $C_{bl}BF$ measurements were performed on 37 postmenopausal women. After the first BBF measurements, 13 and 14 subjects received 0.625 mg and 1.25 mg of CEE per day, respectively, continually for 21 days (low-dose and high-dose groups, respectively). The remaining 10 did not receive CEE (control group). The second BBF measurements were performed on all 37 subjects 21 days after the first measurements. The mean CBF and $C_{bl}BF$ values in the low-dose group were significantly increased during ERT ($p<0.001$), and their percent increases (mean±SD) were 7.1±4.6% and 6.8± 4.9%, respectively. The mean CBF and $C_{bl}BF$ values in the high-dose group were significantly increased during ERT ($p<0.001$), and their mean percent increases were 7.9±5.8% and 6.3±4.7%, respectively. The control group did not show any significant changes during ERT in either the mean CBF or $C_{bl}BF$.

There were no significant differences between low-dose and high-dose groups in either the mean CBF or $C_{bl}BF$.

5.2.1.2 Long-Term ERT/HRT

After the 2nd BBF measurements, 14 out of the 37 patients received HRT for more than one year (long-term ERT/HRT). The content of our HRT is as follows: the patients received 0.625–1.25 mg of CEE from days 1–25 and 5–10 mg of medroxyprogesterone acetate (MPA) from days 14–25 of estrogen treatment, followed by a pause of 7 days (sequential CEE and MPA therapy).

At the mean 13.7±0.5 months after the first BBF measurements, the third BBF measurements were performed on 14 subjects, in the same way as described in Sect. 5.2.1.1. Both the mean CBF and $C_{bl}BF$ were significantly increased during ERT ($p<0.001$ and $p<0.01$, respectively) compared to those of the first BBF measurements, and the mean percent increases were 6.2±5.7% and 6.1±5.8%, respectively. There were no significant differences between the short-term ERT groups and the long-term ERT/HRT group in either the CBF change or the $C_{bl}BF$ change.

5.2.2 Study 2. Effect of Sequential CEE and MPA Therapy on CBF and $C_{bl}BF$

After the first BBF measurements, another 11 naturally menopausal women received sequential CEE and MPA therapy as described above. On the 21st day of treatment, the second BBF measurements were performed on all 11 subjects. Both the mean CBF and $C_{bl}BF$ were significantly increased during sequential CEE and MPA therapy, and the mean percent increases were 7.2±5.6% and 6.8±3.9%, respectively.

These results suggest that short- and long-term ERT and sequential CEE and MPA therapy significantly increase cerebral and cerebellar blood flows to the same degree in postmenopausal women.

5.2.3 Discussion

The mechanisms by which estrogen causes systemic vasodilation have not been clarified sufficiently, but the following possible mechanisms may be considered. Estrogen affects endothelium-derived vasodilators such as prostacyclin (Mueck et al. 1996) and nitric oxide (Rosselli et al. 1995). Another possible mechanism is that estrogen exerts a direct effect on arterial tone because vascular smooth muscle cells possess estrogen receptors and respond to estrogen (Orimo et al. 1993). Furthermore, estrogen shows calcium-antagonistic properties through blocking of voltage-sensitive calcium channels in human aortic smooth muscle (Mueck et al. 1995).

The mean percent increases in both the CBF and C_{bl}BF were much bigger in young postmenopausal women than in both groups of naturally postmenopausal women. The reasons why the percent increases in young postmenopausal women were bigger have not been elucidated completely, but the following possible reasons may be considered. Firstly, the CBF in both men and women decreases with age (Rodriguez et al. 1988). The CBF in young postmenopausal women may respond to ERT better than that in older postmenopausal women. Secondly, the BBF measurements in naturally menopausal women were performed with ^{99m}Tc-labeled ECD. Cerebral uptake of ^{99m}Tc-labeled ECD has been shown to be dependent not only on perfusion but also on the metabolic state (Shishido et al. 1995; Jacquier-Salin et al. 1996). Shishido et al. (1995) reported that ^{99m}Tc-labeled ECD uptake was correlated to the cerebral metabolic rate of oxygen as determined with PET, and that the correlation between ^{99m}Tc-labeled ECD uptake and CBF was less significant. Jacquier-Sarlin et al. (1996) pointed out that the presence of cytosolic esterase, which is related to the viability of cells, is essential for the retention of ^{99m}Tc-labeled ECD in brain tissue. Thirdly, the first BBF measurements were performed on young postmenopausal women four weeks after HRT had been discontinued. In contrast, most of the women with natural menopause received ERT/HRT for the first time. The BBF may have been decreased after four weeks of discontinuation of HRT in young postmenopausal women.

5.2.4 Conclusion

The CBF and C_{bl}BF were significantly increased by three weeks of ERT in women with natural menopause. There were no significant differences between the two doses (0.625 mg/day and 1.25 mg /day) in the percent increases in either CBF or $C_{bl}BF$. The CBF and $C_{bl}BF$ were also significantly increased by long-term (for more than one year) ERT/HRT. The percent increases in the CBF and $C_{bl}BF$ by long-term ERT/HRT were as high as those obtained by three weeks of ERT. Furthermore, the percent increases in the CBF and $C_{bl}BF$ by sequential CEE and MPA therapy were also as high as those by CEE alone. However, changes in the CBF and $C_{bl}BF$ during a combined, continuous CEE and MPA therapy remains to be determined.

5.3 HRT for Dementia of the Alzheimer Type (DAT)

There are many rational justifications for the use of ERT for dementia of the Alzheimer type (DAT). We already described the effects of estrogen on cholinergic neurons and the hippocampus in Sect. 5.1, Estrogen and Memory (Table 2). Other possible rational grounds for HRT use for DAT are as follows: Estrogen promotes the breakdown of amyloid precursor protein (APP) to fragments less likely to accumulate as β-amyloid (Jaffe et al. 1994). APP is expressed during neuronal injury in AD (Regland and Gottfries 1992). In contrast, NGF and BDNF, stimulated by estrogen (Gibbs et al. 1994; Singh et al. 1995), may facilitate the repair of neuronal injury. Estrogen also increases the cerebral uptake and utilization of glucose (Bishop and Simpkins 1992). ERT improves brain blood flow, as described in Sect. 5.1, HRT and Brain Blood Flow. ERT also improves the symptoms of depression in climacteric women (Gerdes et al. 1982). Improvement in depressive symptoms may result in improvement in the cognitive performance of DAT patients. Finally, epidemiological studies on ERT and the risk for DAT (Henderson et al. 1994; Paganini-Hill and Henderson 1994; Tang et al. 1996; Kawas et al. 1997; Baldereschi et al. 1998) revealed that ERT reduces the risk for DAT in postmenopausal women. These rational bases of ERT for HRT are summarized in Table 2. Reports on HRT for female patients with DAT are divided into three parts, and are described below.

Table 2. Rational grounds for the use of estrogen replacement therapy (ERT) for dementia of the Alzheimer type (DAT)

1. Effects on cholinergic neurons
Estrogen increases the activity of choline acetyltransferase (ChAT) and the production of nerve growth factor (NGF) and brain-derived neurotrophic factor (BDNF) receptors.

2. Effects on hippocampus
Estrogen influences the neuronal structure and function of the hippocampus.

3. Attenuation of neuronal injury and facilitation of neuronal repair
Estrogen promotes the breakdown of amyloid precursor protein (APP) to fragments less likely to accumulate as β-amyloid. APP is expressed during neuronal injury in DAT. In contrast, NGF and BDNF stimulated by estrogen may facilitate the repair of neuronal injury.

4. Stimulation of cerebral uptake and utilization of glucose
Estrogen increases the cerebral uptake and utilization of glucose, and glucose transport is dependent on estrogen.

5. Effects on brain blood flow
ERT increases cerebral, cerebellar, and regional cerebral blood flow in postmenopausal women.

6. Effects on memory
Clinical studies show that ERT improves memory in postmenopausal women.

7. Antidepressant effects of estrogen
ERT improves the symptoms of depression in postmenopausal women; this secondarily improves cognitive function in female patients with DAT.

8. Epidemiological studies on ERT and risk for DAT
ERT reduces the risk for DAT in postmenopausal women.

5.3.1 Effects of Short-Term High-Dose ERT on Cognitive Function, Dementia Symptoms, rCBF, and EEG Activity in DAT Patients

This study (Ohkura et al. 1994a) was designed to investigate the therapeutic efficacy of short-term high-dose ERT in female patients with DAT. Fifteen DAT patients with a mean age of (mean±SE) 71.9±2.4 years were orally treated with 0.625 mg of CEE twice a day for 6 weeks. Of the 15 DAT patients, 4 were diagnosed as mild, 7 as moderate and 4 as severe. The effects of estrogen on DAT patients were evaluated by psychometric assessments, behavior rating scales, regional cerebral blood flow (rCBF) measurement, and quantitative EEG analysis. Psychometric assessments consisted of mini-mental state examination (MMSE) and Hasegawa dementia scale (HDS). Dementia symptoms were evaluated by the GBS scale (GBSS) and the Hamilton depression rating scale (HDRS). During ERT, the mean MMSE score (mean±SE) was significantly increased from 11.6±1.9 to 13.2±2.0 at 3 weeks ($p<0.01$) and 13.8±2.0 at 6 weeks ($p<0.001$). The mean HDS score was significantly increased from 8.6±2.1 to 11.5±2.3 at 3 weeks ($p<0.001$) and 11.6±2.6 at 6 weeks($p<0.01$). Significant improvements in the mean scores of the GBSS and HDRS were also observed in the estrogen-treated group, but not in the untreated control group with a mean age of 71.2±2.5 years ($n=15$). The rCBF was measured by SPECT and [^{123}I]IMP. ERT increased the rCBF significantly in the lower frontal region ($p<0.01$) and primary motor area ($p<0.02$) of the right hemisphere. The mean absolute power delta band values in both left and right frontal EEG (Fp_1 and Fp_2) ($p<0.01$) and theta 1 band values in Fp_2 ($p<0.05$) were significantly decreased during ERT. From this it is inferred that ERT significantly improves cognitive function, dementia symptoms, regional cerebral blood flow, and EEG activity in female patients with DAT.

5.3.2 Effect of Long-Term Low-Dose ERT on Cognitive Function in DAT Patients

This study (Ohkura et al. 1994a) was designed to investigate the effect of long-term low-dose ERT on cognitive function in female patients

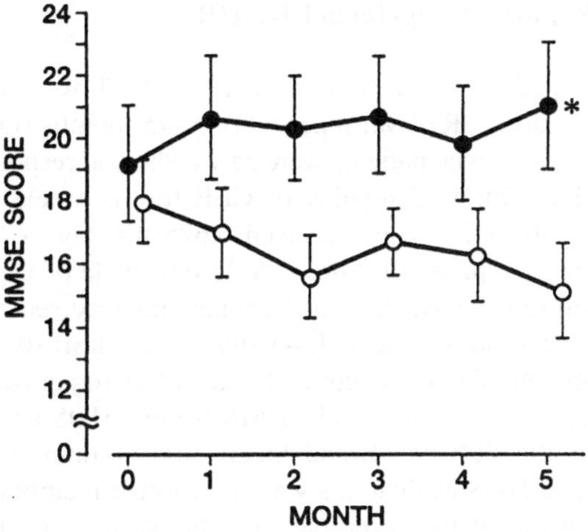

Fig 5. Mini-mental state examination (MMSE) scores for the estrogen-treated group (*filled circles*) and the untreated control group (*open circles*). There was a significant group × time interaction for the MMSE ($F_{5,90}$=5.815, p=0.0001). *Vertical bars* indicate the mean±SE; *$p<0.05$ vs. control (from Ohkura et al. 1994b)

with mild to moderate Alzheimer's disease (AD). Ten patients with AD received 0.625 mg/day of CEE for 21 days, followed by a pause of 7 days. The ERT cycle consisting of 28 days was repeated for 5 months in the 10 patients (EST group). Another 10 patients with AD were recruited as an untreated control group (CON group). Psychometric assessments with MMSE were performed once a month. There was a significant group × time interaction for the MMSE (Fig. 5). The mean total MMSE score in the 5th month was significantly higher in the EST group than in the CON group ($p<0.05$). A significant difference between the two groups in each mean score for each question of the MMSE performed in the 5th month was observed in the questions regarding orientation in time ($p<0.05$) and recalling three objects ($p<0.01$). These results suggest that low-dose ERT may either improve cognitive function or slow the rate of cognitive decline in patients with mild to moderate Alzheimer disease (Fig. 5).

5.3.3 Case Reports: Long-Term ERT/HRT

Seven female patients with mild to moderate DAT were treated with long-term, low-dose ERT over a period of 5–45 months (Ohkura et al. 1995b). Five of the seven patients were cases who had responded well to short-term ERT with 1.25 mg/day of CEE for six weeks. The seven patients, aged 56 to 77 years, received 0.625 mg/day of CEE for 21 days, followed by a pause of 7 days. A 28-day cycle of low-dose ERT was performed repeatedly. Four of these patients received 5 mg/day of medroxyprogesterone acetate (MPA) during the last 10–12 days of estrogen treatment. The therapeutic efficacy of estrogen was evaluated by psychometric assessments such as MMSE and HDS and a behavior rating scale of the Gottfries–Brane–Steen geriatric rating scale (GBS). The MMSE and HDS evaluations were performed mainly once in 2–4 weeks. In four out of the seven patients, the MMSE and HDS scores were elevated above the pretreatment levels during ERT. The termination of ERT resulted in a decrease in both scores. Furthermore, the GBS score and daily activities of the same four patients were improved during ERT. In the four patients, cognitive function was markedly improved throughout the treatment period, while the other two patients responded moderately well and another patient did not respond at all. These observations suggest that long-term, low-dose ERT improves cognitive function, dementia, symptoms, and daily activities in women with mild to moderate DAT. However, the supplemental treatment with MPA seems to have an unfavorable effect on the dementia symptoms and the daily activities of the aged DAT patients.

Fig. 6 shows a case (H.M.) for whom long-term ERT/HRT for four years was very effective (Ohkura et al. 1995b).

When the patient was 69 years old, her family recognized her conspicuous forgetfulness and her repeating of the same words and phrases. During the next four years, the family reported that the patient was rapidly losing memory. She first visited the outpatient clinic of psychiatry of the Tokyo Metropolitan Tama Geriatric Hospital on March 22, 1990,when she was 76 years old. the MMSE and HDS scores of the patient at the outpatient clinic were between 10 and 11, and between 6 and 8.5, respectively (Fig. 6). The GBS score was 53. The diagnosis was moderate DAT. The serum estradiol level was less than 10 pg/ml and the

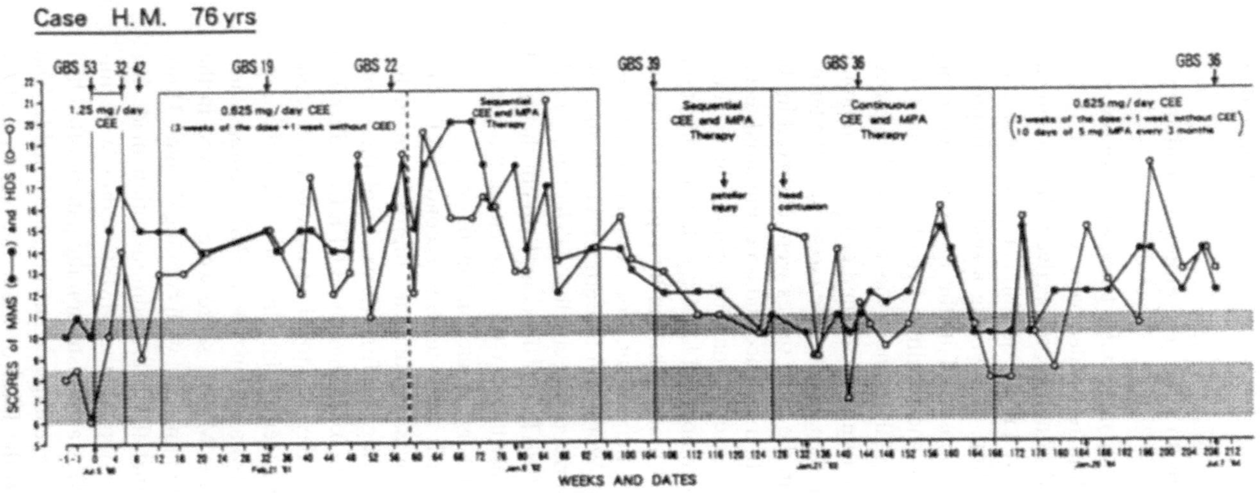

Fig 6. Changes in the scores of the MMSE, HDS, and GBS scale during ERT/HRT in a moderate DAT patient (patient: H.M.). The age on the diagram indicates the patient age when ERT was first initiated (from Ohkura et al. 1995b)

patient had no contraindications to ERT. Therefore, the patient was selected as a case for ERT.

The patient orally received 0.625 mg of CEE twice a day for 6 weeks when she was 76 years old. She responded very well to ERT. The MMSE and HDS scores reached 17 and 14, respectively. Improvements in the MMSE scores were observed in the questions regarding orientation in time and space, calculation, and writing a sentence. Improvements in the HDS scores were observed in the questions regarding orientation in space, calculation, and recalling five objects. The GBS score was decreased to 32 during ERT. There was a marked improvement in symptoms of intellectual impairment, and different symptoms common in dementia. However, the score returned to 42 after the termination of ERT. Improvements in the GBS scores were observed in the items regarding impaired orientation in space and time, impaired recent and distant memory, impaired concentration, absentmindedness, long-windedness, emotional blunting, emotional lability, reduced motivation, irritability, anxiety, and reduced mood. The family reported improvements in the daily activities of the patient. For example, the patient talked about her experience after World War II. Before ERT, she had told the family only about the memories of her childhood. The family made a request for the continuation of ERT after the completion of the initial six-week treatment with 1.25 mg/day of CEE. Therefore, the patient received long-term, low-dose ERT as depicted in Fig. 6.

The patient again responded well to the second ERT with 0.625 mg/day of CEE. There was a moderate improvement in the MMSE scores of the patient during the first 45 weeks of the second ERT. We continued treatment with 0.625 mg/day of CEE from days 1 to 21 and 5 mg/day of MPA from days 12 to 21 of estrogen treatment, followed by a pause of 7 days (sequential CEE and MPA therapy), which brought a drastic improvement in the MMSE score, which reached 20. Improvements in the MMSE score were observed in the questions regarding orientation in time and with copying designs. However, her family complained that the MPA administration made her physical condition worse, even though the MMSE scores were improved. For example, during MPA administration the patient often had an urge to go to her home in the evening and she became irritable. The MMSE score had decreased to the level of moderate response at 60 weeks after the initiation of the second ERT, even though ERT was continued. The HDS scores reached the previous peak level soon after the second ERT. However, in the HDS scores, the patient showed fluctuating responses after the first 22 weeks of the second ERT. ERT was discontinued in order to evaluate the therapeutic efficacy of estrogen. The MMSE scores decreased further after the discontinuation of ERT. The family reported that the patient's daily activities had deteriorated more than might be expected from the decrease in the test score. Therefore, we abandoned an untreated control study and began ERT again after 10 weeks of discontinuation.

One hundred twenty-seven weeks after the initiation of the first ERT, a continuous regimen of estrogen and low-dose progestogen (continuous, combined CEE and MPA therapy: daily 0.625 mg of CEE coupled with 2.5 mg of MPA) was adopted to reduce withdrawal bleeding. But the resumption of ERT (the third ERT) could not restore the patient's daily activities and the test scores to the levels before the discontinuation of ERT. To make matters worse, on October 10, 1992, she sprained her right knee. As she was unable to walk, she could not go out for 6 weeks (118 weeks after the initiation of the first ERT). Furthermore, she fell down and bumped her head on December 30 (129 weeks). These accidents caused a further decrease in the MMSE scores. At 53 weeks after the initiation of the third ERT, both MMSE and HDS scores were increased transiently to the levels before the third ERT, followed by an abrupt decrease in both scores to pretreatment levels. The family reported that her poor physical condition was not provoked by a continuous regimen of CEE and low-dose MPA. However, they did not observe any more improvement in her daily life. She obtained continuous CEE and MPA therapy until September 16, 1993 (168 weeks after the first ERT) and then the 28-day cycle of ERT was resumed, in which withdrawal bleeding was induced following the administration of 5 mg/day of MPA during the last 10 days of estrogen treatment every 3 months. During the 28-day-cycle of ERT, the MMSE and HDS scores were fairly good (Fig. 6). It seemed to us that the cyclic administration of CEE and the reduced administration of MPA might have had a favorable influence on the patient's physical condition and test scores.

Two GBS scores were evaluated during the second ERT; they were 19 and 22. The GBS scores during the second ERT had improved further in the items regarding the following: deficiency of spontaneous activity, motor insufficiency in managing personal hygiene, impaired orientation in time, impaired personal orientation, impaired distant memory, absent-mindedness, and reduced mood. However, the termination of ERT brought the GBS score back to 39. Marked deterioration was observed in intellectual function. The GBS score was 36 at 143 weeks after the initiation of the first ERT. In the last assessment on July 7, 1994 (209 weeks after the first ERT), the MMSE and HDS scores were 12 and 13, respectively. The GBS score was 36. Long-term, low-dose ERT was continued for 45 months, except for 10 weeks of discontinuation, and 4 years have just passed on July 7, 1994, since the first ERT was initiated. It seemed likely that ERT in this case might arrest the course of DAT at least from the viewpoint of the MMSE, HDS, and GBS scores. Withdrawal bleeding was frequently observed during the 7-day pause of both unopposed ERT and sequential CEE and MPA therapy. A small quantity of withdrawal bleeding was sporadically observed during the continuous, combined CEE and MPA therapy. During the unopposed ERT after the continuous, combined CEE and MPA therapy, withdrawal bleeding was scarcely observed, but the administration of MPA along with CEE induced withdrawal bleeding. On the whole, long-term ERT was very effective in this case. However, 5 mg/day of MPA administration along with CEE made the patient's physical condition worse.

5.3.4 Conclusion

ERT for 6 weeks was associated with significant improvements not only in cognitive function and dementia symptoms, but also in regional cerebral blood flow and EEG activities in female patients with DAT. Long-term ERT is also effective for the improvements in cognitive function and clinical symptoms of DAT patients. But the administration of MPA caused unfavorable side effects in some DAT patients. A better application method for progestogen (e.g., vaginal administration or an intrauterine device) still needs to be found.

Acknowledgements. The studies "Estrogen and Brain Blood Flow" (Sect. 5.2) and "HRT for Dementia of the Alzheimer Type" (Sect. 5.3) were supported by Grants-in-Aid for Scientific Research from the Ministry of Education, Science, and Culture (No. 08671920 and No. 04670707, respectively).

References

Baldereschi M, Carlo AD, Lepore V, Maggi S, Grigoletto F, Scarlato G, Amaducci L (1998) Estrogen-replacement therapy and Alzheimer's disease in the Italian Longitudinal Study on Aging. Neurology 50:996–1002

Belchetz PE (1994) Hormonal treatment of postmenopausal women. N Engl J Med 330:1062–1071

Bishop J, Simpkins JW (1992) Role of estrogens in peripheral and cerebral glucose utilization. Rev Neurosci 3:121–137

Gangar KF, Vyas S, Whitehead M, Crook D, Meire H, Campbell S (1991) Pulsatility index in internal carotid artery in relation to transdermal oestradiol and time since menopause. Lancet 338:839–842

Gerdes LC, Sonnendecker EWW, Polakow ES (1982) Psychological changes effected by estrogen–progestogen and clonidine treatment in climacteric women. Am J Obstet Gynecol 142:98–104

Gibbs RB, Wu D, Hersh LB, Pfaff DW (1994) Effects of estrogen replacement on the relative levels of choline acetyltransferase, trkA, and nerve growth factor messenger RNAs in the basal forebrain and hippocampal formation of adult rats. Exp Neurol 129:70–80

Gould E, Woolley CS, Frankfurt M, McEwen BS (1990) Gonadal steroids regulate dendritic spine density in hippocampal pyramidal cells in adulthood. J Neurosci 10:1286–1291

Henderson VW, Paganini-Hill A, Emanuel CK, Dunn ME, Buckwalter JG (1994) Estrogen replacement therapy in older women: comparisons between Alzeheimer's disease cases and nondemented control subjects. Arch Neurol 51:896–900

Jacquier-Sarlin M, Polla B, Siosman D (1996) Cellular basis of ECD brain retention. J Nucl Med 37:1694–1697

Jaffe AB, Toran-Allerand CD, Greengard P, Gandy SE (1994) Estrogen regulates metabolism of Alzheimer amyloid β precursor protein. J Biol Chem 269:13065–13068

Kampen DL, Sherwin BB (1994) Estrogen use and verbal memory in healthy postmenopausal women. Obstet Gynecol 83:979–983

Kawas C, Resnick S, Morrison A, Brookmeyer R, Corrada M, Zonderman A, Bacal C, Donnell Lingle D, Metter E (1997) A prospective study of estrogen replacement therapy and the risk of developing Alzheimer's disease: the Baltimore Longitudinal Study of Aging. Neurology 48:1517–1521

Lehtovirta P (1974) Haemodynamic effects of combined oestrogen- progestogen oral contraceptives. Obstet Gynecol Br Commonwealth 81:517–525

Luine VN (1985) Estradiol increases choline acetyltransferase activity in specific basal forebrain nuclei and projection areas of female rats. Exp Neurol 89:484–490

Mashini IS, Albazzaz SJ, Fadel HE, Abdulla AM, Hadi HA, Harp R, Devoe LD (1987) Serial noninvasive evaluation of cardiovascular hemodynamics during pregnancy. Am J Obstet Gynecol 156:1208–1213

Matsuda H, Yagishita A, Tsuji S, Hisada K (1995) A quantitative approach to technetium-99m ethylcysteinate dimer: a comparison with technetium-99m hexamethylpropylene amine oxide. Eur J Nucl Med 22:633–637

Mueck AO, Seeger H, Hanke H, Lippert TH (1995) Calcium-antagonistic effect of estradiol compared with nifedipine, verapamil, and diltiazem – in vitro investigations in human aortic smooth muscle cells [Abstract]. Menopause 2:262–263

Mueck AO, Seeger H, Korte K, Lippert TH (1996) The effect of 17β -estradiol and endothelin 1 on prostacycline and thromboxane production in human endothelial cell cultures. Clin Exp Obstet Gynecol 20:203–206

Ohkura T, Isse K, Akazawa K, Hamamoto M, Yaoi Y, Hagino N (1994a) Evaluation of estrogen treatment in female patients with dementia of the Alzheimer type. Endocr J 41(4):361–371

Ohkura, Isse K , Akazawa K, Hamamoto M, Yaoi Y, Hagino N (1994b) Low-dose estrogen replacement therapy for Alzheimer disease in women. Menopause 3:125–130

Ohkura T, Teshima Y, Isse K, Matsuda H, Inoue T, Sakai Y, Iwasaki N, Yaoi Y (1995a) Estrogen increases cerebral and cerebellar blood flows in postmenopausal women. Menopause 2(1):13–18

Ohkura T, Isse K, Akazawa K, Hamamoto M, Yaoi Y, Hagino N (1995b) Long-term estrogen replacement therapy in female patients with dementia of the Alzheimer type: 7 case reports. Dementia 6:99–107

Ohkura T, Matsuda H, Iwasaki N, Teshima Y, Natsui S, Isse K, Inaba N, Yaoi Y (1996) Effect of estrogen on regional cerebral blood flow in postmenopausal women. J Jpn Menopause Soc 4(2):254–261

Orimo A, Inoue S, Ikegami A, Hosoi T, Akishita M, Ouchi Y, Muramatsu M, Orimo H (1993) Vascular smooth muscle cells as target for estrogen. Biochem Biophys Res Commun 195:730–736

Paganini-Hill A, Henderson VW (1994) Estrogen deficiency and risk of Alzheimer's disease in women. Am J Epidemiol 140:256–261

Penotti M, Nencioni T, Gabrielli L, Farina M, Castiglioni E, Polvani F (1993) Blood flow variations in internal carotid and middle cerebral arteries induced by postmenopausal hormone replacement therapy. Am J Obstet Gynecol 169:1226–1232

Phillips SM, Sherwin BB (1992) Effects of estrogen on memory function in surgically menopausal women. Psychoneuroendocrinology 17:485–495

Regland B, Gottfries CG (1992) The role of amyloid beta protein in Alzheimer's disease. Lancet 340:467–469

Rodriguez G, Warkentin S, Risberg J, Rosadini G (1988) Sex differences in regional cerebral blood flow. J Cereb Blood Flow Metab 8:783–789

Rosselli M, Imthurn B, Keller PJ, Jackson EK, Dubey RK (1995) Circulating nitric oxide (nitrite/nitrate) levels in postmenopausal women substituted with 17β-estradiol and norethisterone acetate; a two-year follow-up study. Hypertension 25:848–853

Shishido F, Uemura K, Inugami A, Ogawa T, Fujita H, Shimosegawa E, Nagata K (1995) Discrepant ^{99m}Tc-ECD images of CBF in patients with subacute infarction: a comparison of CBF, CMRO2 and ^{99m}Tc-HMPAO imaging. Ann Nucl Med 9:161–166

Singh M, Meyer EM, Simpkins JW (1995) The effects of ovariectomy and estradiol replacement on brain-derived neurotrophic factor messenger ribonucleic acid expression in cortical and hippocampal brain regions of female Sprague-Dawley rat. Endocrinology 136:2320–2324

Tang M-X, Jacobs D, Stern Y, Marder K, Schofield P, Gurland B, Andrews H, Mayeux R (1996) Effect of oestrogen during menopause on risk and age at onset of Alzheimer's disease. Lancet 348:429–432

Veille JV, Morton MJ, Burry K, Nemethy M, Speroff L (1986) Estradiol and hemodynamics during ovulation induction. J Clin Endocrinol Metab 63:721–724

Woolley CS, Weiland NG, McEwen BS, Schwartzkroin PA (1997) Estradiol increases the sensitivity of hippocampal CA1 pyramidal cells to NMDA receptor-mediated synaptic input: correlation with dendritic spine density J. Neurosci 17:1848–1859

6 Hormone Replacement Therapy and Lipids

H. Honjo, M. Urabe, K. Tanaka, T. Kashiwagi, T. Okubo,
H. Tsuchiya, K. Iwasa, N. Kikuchi, and T. Yamamoto

6.1 Introduction

In 1997, the average life span for Japanese women was 83.82 years, the highest worldwide average for 13 years. However, sometimes the faces of Japanese women don't look so well. Their quality of life is also very important. The second most common cause of death in women is cerebrovascular disease, and the third is heart disease. Both of these conditions are closely associated with arterial sclerosis (Table 1).

The Population Development Statistics for 1995 by the Japanese Ministry of Health and Welfare (1996) show that mortality due to ischemic heart disease in those aged 25–60 years is approximately four times higher in men than in women. But the male–female gap narrows in the 60-plus age range (i.e., approximately 10 years after menopause), and after 75 years of age the two groups do not significantly differ

Table 1. Estimated rate of leading causes of death (% of possible causes of death for those born in 1997) (Japanese Ministry of Health and Welfare, 1998)

Ranking	Males	(%)	Females	(%)
1st	Cancer	29.43	Cancer	19.54
2nd	Heart disease	14.78	Cerebrovascular disease	18.84
3rd	Cerebrovascular disease	14.63	Heart disease	18.27

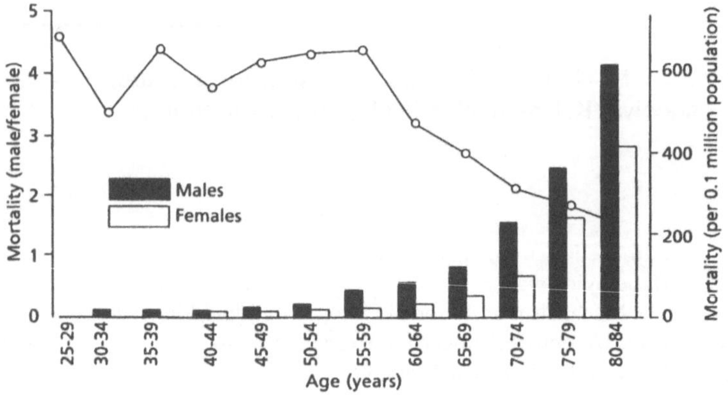

Fig. 1. Age-related mortality due to ischemic heart disease in males and females in 1995

(Fig. 1). In other words, postmenopausal women are susceptible to ischemic heart disease, largely due to the reduction in estrogen following the menopause (Honjo et al. 1994).

6.2 Hyperlipidemia

The mean level of serum cholesterol of women in the USA has gradually lowered, but that of Japanese women has gradually increased. Finally, by 1990, the latter became slightly higher than the former.

Hyperlipidemia, the most common cause of arterial sclerosis, is significantly increased in postmenopausal women. Serum total cholesterol (TC) above 220 mg/dl (the critical level, i.e., the level at which

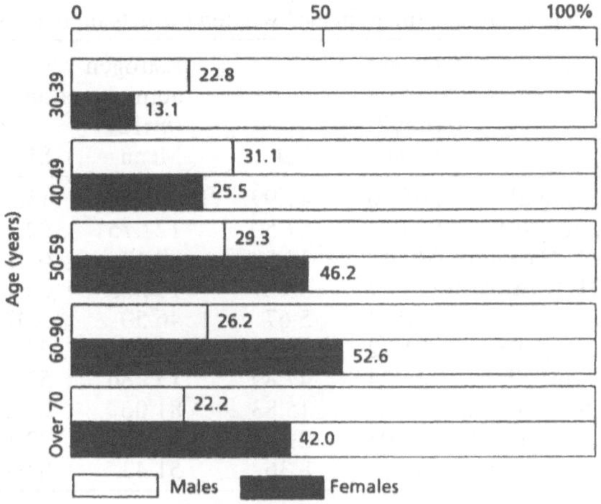

Fig. 2. Percentage of men and women with serum TC values above 220 mg/dl

treatment of hyperlipidemia should begin, according to the Consensus Conference of the Japanese Arterial Sclerosis Society, winter 1987) is seen in 46.2% of women in their fifties, and 52.6% in their sixties. Both figures are considerably higher than those in age-matched men (Fig. 2) (Japanese Ministry of Health and Welfare, 1993). The mean menopausal age is 50.0 year old in the Kinki district, Japan (Hirota et al. 1995). The TC levels of postmenopausal climacteric women aged 50–52 years are significantly higher than those of premenopausal climacteric women (Honjo 1991). The relationship between menopause, reduction in estrogen, and increase in TC is an important factor in the development of hypercholesterolemia.

Pre- and postmenopausal women were compared in a study of one group of women on estrogen replacement therapy (for treatment of climacteric symptoms and osteoporosis) and a control group (Honjo et al. 1992). The high-density lipoprotein cholesterol (HDL-C) was found to be raised and the low-density lipoprotein cholesterol (LDL-C) was reduced in the postmenopausal group on estrogen replacement therapy (Table 2). A significant decrease in TC can generally be demonstrated after estrogen administration, although this was not demonstrable in this study.

Table 2. Serum lipids in pre- and postmenopausal women (effects on extrinsic estrogen)

		Controls (n=62)		Estrogen administration (n=12)		Significance[a]
		Mean	SD	Mean	SD	
Premenopause	TC (mg/dl)	207.16	41.99	195.50	15.02	n.s.
	TG (mg/dl)	98.09	51.50	127.75	92.30	n.s.
	HDL-C (mg/dl)	61.91	14.58	69.13	6.45	n.s.
	LDL-C (mg/dl)	124.47	38.27	100.83	25.72	n.s.
	Age	43.66	5.67	46.50	5.07	n.s.
Postmenopause	TC (mg/dl)	231.72	29.48	226.14	33.19	n.s.
	TG (mg/dl)	114.24	47.89	133.86	53.29	n.s.
	HDL-C (mg/dl)	61.82	15.83	81.06	25.00	p<0.05
	LDL-C (mg/dl)	147.05	29.40	118.31	22.17	p<0.05
	Age	58.00	8.36	51.43	8.94	n.s.

TG, triglyceride.
[a]Student's t-test.

6.3 The Mechanism of Action of Estrogen on Lipid Metabolism

The effects of estrogen on hepatic triglyceride lipase (HTGL) and lipoprotein lipase (LPL) are compared in Fig. 3. Estrogen replacement therapy in postmenopausal women suppresses HTGL remarkably, but does not affect LPL significantly (Urabe et al. 1996).

Figure 4 shows the effects of estrogen on enzymes and lipids (Honjo et al. 1994). Low- or medium-dose estrogen reduces TC and raises HDL-C. HTGL accelerates the catabolic conversion of TG-rich (TG, triglyceride) HDL_2 to TG-poor HDL_3. The suppression and deficit of HTGL may increase HDL_2. HTGL also induces hepatic uptake of HDL.

The suppression and decrease of HTGL may encourage the increase in HDL. Estrogen may stimulate intestinal and/or hepatic production of apolipoprotein A-I (apoA-I) and increase HDL (the main component of which is apoA-I). Synergy of these estrogenic effects may induce the significant increase in HDL-C. HTGL also accelerates the conversion of intermediate-density lipoprotein (IDL) to LDL (-C). The suppression of HTDL by estrogen results in the reduction of LDL (-C) as well as serum TC. In addition, estrogen increases hepatic LDL receptors and acceler-

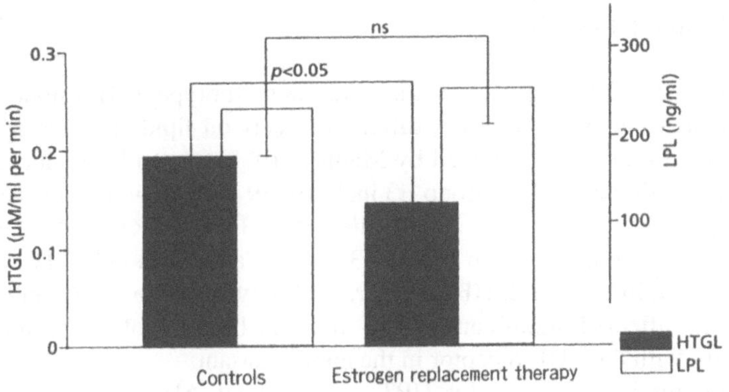

Fig. 3. Estrogen replacement therapy in postmenopausal women significantly suppresses hepatic triglyceride lipase (HTGL), yet has no effect on lipoprotein lipase (LPL)

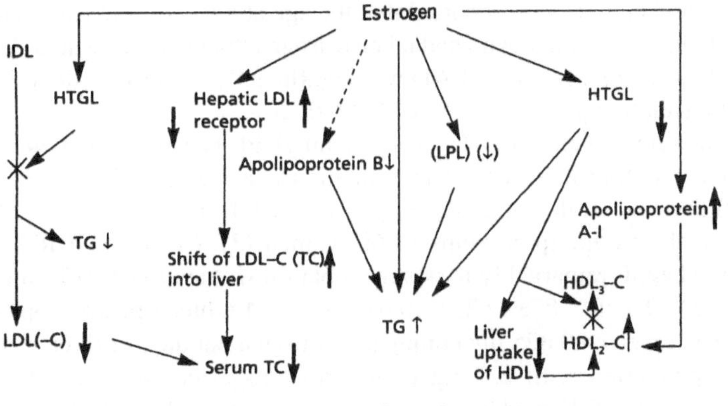

Fig. 4. Beneficial effects of low- to medium-dose estrogen on enzymes and lipids

ates the shift of serum LDL-C into the liver, leading to the decrease in serum LDL-C (and TC).

Low- or medium-dose estrogen has beneficial effects on lipids by the above mechanism. However, a high dose may lead to an increase in serum triglyceride (TG) and have an unfavorable effect, by causing an increase in hepatic production of TG and inhibition of LPL.

6.3.1 Apolipoprotein E

Apolipoprotein E (apoE) has three variants (subtypes). Hormone replacement therapy (HRT) has different effects on lipids in each apoE subtype. In a study conducted by Mahley (1988), group E2 included 6 women with allele ε2/3. Group E3 included 39 women with allele ε3/3. Group E4 included 7 women with allele ε3/4. Under HRT, LDL-C was suppressed significantly in group E3 and E4 after 3 months. TC was suppressed in group E3. HDL-C increased only in group E3. Group E2 was not affected significantly. This may be because of low binding activity with the LDL receptor in the apoE2 variant.

Suppression of apoE by HRT may prevent Alzheimer's disease (Honjo et al. 1995, Honjo et al. 1997, Honjo et al. 1998). Many epidemiological studies of HRT and Alzheimer's disease (AD) have been performed recently. Paganini-Hill and Henderson (1996) reported that those treated with estrogen replacement therapy have a lower risk of AD. Tang et al. (1996) reported that the age at onset of AD was significantly later in women who had taken estrogen than in those who did not, and that the relative risk of AD was significantly reduced in the former group (relative risk 0.40; 95% CI 0.22–0.85).

The senile plaques in AD consist of amyloid. The main component of amyloid is β-protein, whose precipitation is accelerated by apoE. It is well known that the ε4 allele of apoE is a risk factor for late-onset AD. We analyzed apolipoproteins in 68 women (37–67 years). HRT suppressed apoE remarkably in postmenopausal women (3.41±0.75 mg/dl, mean±S.D.; cf. 4.87±1.63, $p<0.01$, Fig. 5). Eighteen postmenopausal women received 0.625 mg conjugated estrogen per day for three weeks, in combination with 2.5 mg of medroxyprogesterone acetate for 10 days; this was followed by a one-week pause, after which the four-week cycle was repeated. The serum level of apoE decreased significantly during the third and sixth months because of HRT.

A mass screening for AD was performed in a northern town in the Kyoto prefecture. The phenotype of apoE was analyzed. E3/4 was found in 35% of 20 women suffering from AD. E4/4 (1%), E2/4 (1%), and E3/4 (16%) were found in 68 normal women. The serum level of estrone (E_1) was significantly ($p<0.01$) lower in women with AD than in normal women. Obese women produce more E_1 by aromatization of androgen

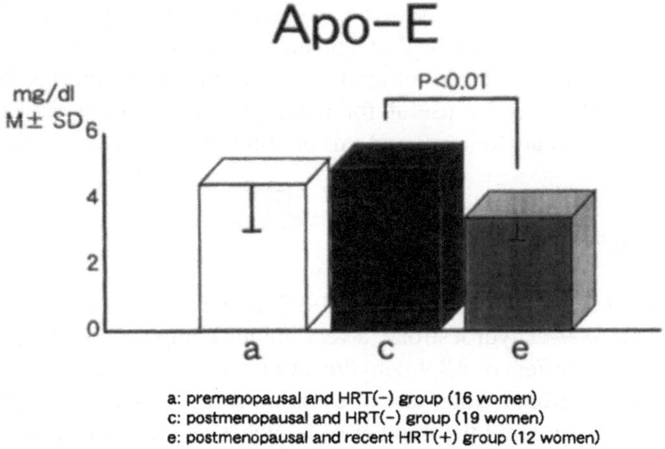

Fig. 5. Apolipoprotein E and HRT: apolipoprotein E was remarkably suppressed by HRT in postmenopausal women

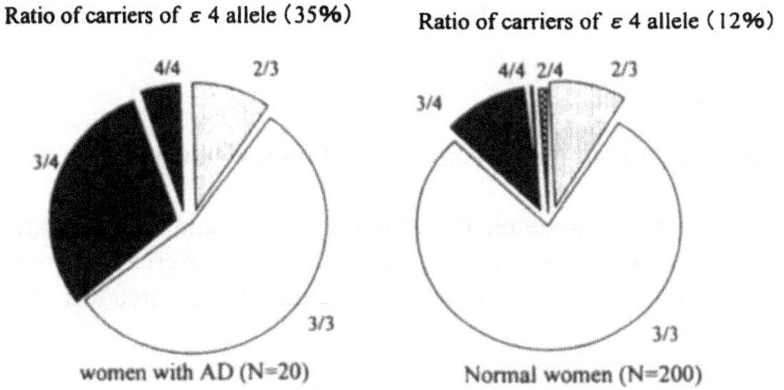

Fig. 6. Genotype of apoE; χ^2 test ($p<0.01$)

in peripheral fatty tissue. The E_1 that is produced may be useful in preventing AD.

The genotype of apoE was analyzed with the PCR-RFLP method. Allele ε2, ε3, and ε4 are the main isoforms of the apoE allele. The apoE genotype consisted of six subtypes: ε2ε2, ε2ε3, ε2ε4, ε3ε3, ε3ε4, and ε4ε4. The ratio of carriers of ε4 was significantly higher in the 20 women with AD than in the 200 women without AD (Fig. 6).

Even in Japan, allele ε4 of apoE is one of the important risk factors for AD. HRT may be useful for these ε4 carriers, even for elderly women without any other symptoms or abnormal laboratory data.

6.3.2 Δ8,9-Dehydroestrone

Brinton et al. (R.D. Brinton et al. 1997, personal communication) reported on Δ8,9-dehydroestrone, a very unique component of conjugated estrogen. The effect of Δ8,9-dehydroestrone on neuronal outgrowth was shown to be stronger than that of estradiol and even that of conjugated estrogen. It increased the number of branches and the branch length of neurons. X-ray analysis showed that its structure differs from those of estrogen and androgen, but in opposite ways. It is possible for Δ8,9-dehydroestrone to have a neutral structure, namely, to be flatter between the B ring and C ring, because of the Δ8,9 double bond. One may use Δ8,9-dehydroestrone as a substance against AD, but with less estrogenic effects.

6.4 Direct Action of Estrogen on Arterial Sclerosis

Inhibition of hyperlipidemia by estrogen prevents and improves arterial sclerosis. Estrogen also has a direct action on arterial sclerosis by inhibiting oxidation and suppressing the growth or spread of smooth muscle cells.

6.4.1 Inhibition of Oxidation

Lipid peroxide degenerates (oxidizes) LDL-C, which is then absorbed by the monocyte, and turned into foam cells in the vascular subendothelial cavity; this leads to atherogenesis in the aorta and eventually to arterial sclerosis. Estrogen, like some other substances, can suppress this oxidation (Fig. 7).

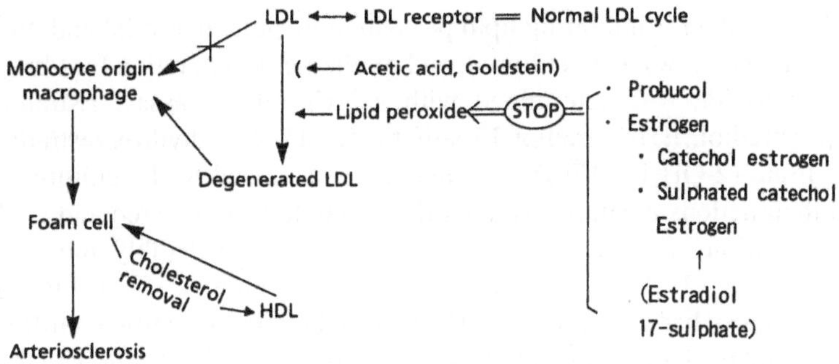

Fig. 7. Lipid peroxidation and sulfated catechol estrogen (2- or 4-hydroxy estradiol 17-sulfate)

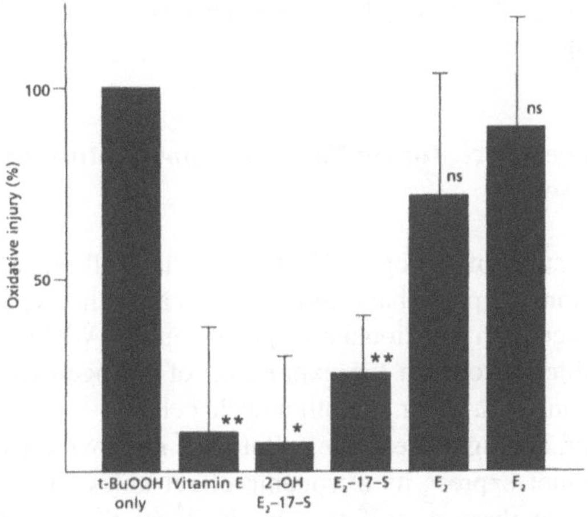

Fig. 8. Analysis of oxidation of human arterial endothelial cells by *t*-BuOOH, using a ^{51}Cr release assay: the addition of 2-OH E_2–17-S suppresses oxidation, as does vitamin E. E_2–17-S also has a suppressive effect. Means±SD; *p<0.01; **p<0.02

As a model for inhibiting lipid peroxidation, human arterial endothelial cells (Krabow) were cultured, and oxidizing degeneration by *t*-butyl hydroperoxide was determined with a ^{51}Cr release assay. Estradiol (17β-estradiol, E_2), estradiol 17-sulfate (E_2–17-S), 2-hydroxyestradiol 17-sulfate (2-OH E_2–17-S), or vitamin E was added to the culture, so that their actions in suppressing oxidation could be compared and studied. Addition of 2-OH E_2–17-S or vitamin E considerably inhibited oxidation; E_2–17-S also showed a significant inhibitory effect (Fig. 8). Estradiol also has a suppressive effect, even though no significant difference could be demonstrated in this study. This study clearly showed that estrogen, especially E_2–17-S and 2-OH E_2–17-S, effectively suppresses oxidation. It is thought that these estrogens may be effective in preventing oxidative stress and in antagonizing atherogenesis as well as hypertension.

Interestingly, Δ8,9-dehydroestrone also strongly inhibits oxidized LDL binding.

6.4.2 Estrogen Receptors in Vascular Smooth Muscle Cells, and Sulfatase

The direct action of estrogen on the vascular cells involves estrogen receptors. Some reports have already indicated that vascular smooth muscle cells contain functional estrogen receptors. We have also studied and been able to confirm the expression of estrogen receptors in the mRNA of human vascular smooth muscle cells.

Moreover, conjugated estrone sulfate, the most widely used estrogen in HRT, cannot express its estrogenic effect unless it is changed into active estrogen through sulfatase in the local tissues. We have also confirmed the expression of sulfatase mRNA in the human vascular smooth muscle cells and, with in situ hybridization, showed that sulfatase is present in it. These results suggest that human vascular smooth muscle cells have sulfatase activity which can convert conjugated estrone sulfate to active estrone and estradiol in local cells; this leads to direct effects on vascular cells such as estrogen receptors.

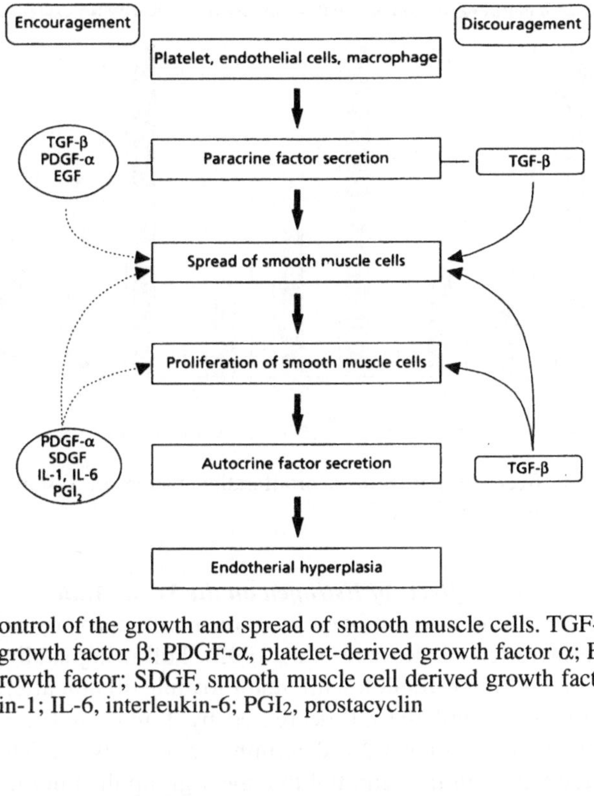

Fig. 9. Control of the growth and spread of smooth muscle cells. TGF-β, transforming growth factor β; PDGF-α, platelet-derived growth factor α; EGF, epidermal growth factor; SDGF, smooth muscle cell derived growth factor; IL-1, interleukin-1; IL-6, interleukin-6; PGI_2, prostacyclin

6.4.2.1 Control of Growth and Spread of Smooth Muscle Cells

Another important mechanism in arterial sclerosis is the progress of the medial vascular smooth muscle cells (spread into the vascular subendothelial cavity→proliferation→arterial sclerosis). It is thought that multiple paracrine or autocrine factors are involved in this process (Fig. 9). Human vascular smooth muscle cells (derived from aorta, Krabow) or human umbilical venous endothelial cells were cultured to investigate the effect of estrogen on mRNA expression in each factor. Estrogen considerably suppressed mRNA expression of platelet-derived growth factor α (PDGF-α), interleukin-1 (IL-1), and interleukin-6 (IL-6), which accelerate the arterial sclerosis progress (Fig. 10). These outcomes suggest that estrogen suppresses the progress of arterial sclerosis in both paracrine and autocrine factors.

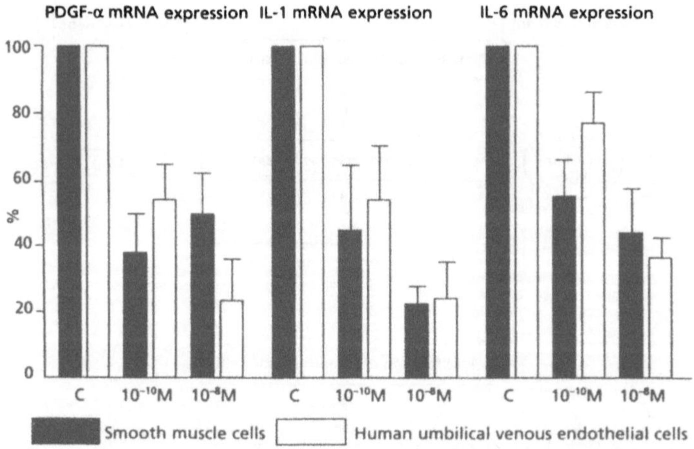

Fig. 10. The effects of estrogen on proliferating factors of vascular smooth muscle cells and endothelial cells

6.4.2.2 Immediate Effects of Estrogen on the Vessel Wall

The immediate effects of estrogen in women with coronary artery disease were investigated (Rosano et al. 1993). Following sublingual administration of 1 mg estradiol, with motor loading 40 min after administration, the time until the ST decreased by 1 mm, and the time of continuous motor loading were determined and analyzed. These time delays were longer in the estradiol treatment group than in the placebo group. An acute/chronic improvement in cerebral blood flow with estrogen replacement therapy was also documented (Ohkura et al. 1995).

6.5 Estrogen Replacement Therapy

Long-term treatment with estrogen is thought to improve hyperlipidemia and arterial sclerosis and to inhibit or improve cardiovascular disease and may prevent AD. Estrogen replacement therapy prolongs the life span by 0.9 years in women with no risk factors and by 2.1 years in women with a history of coronary heart disease (Table 3) (Grady et al. 1992).

Table 3. Net change in life expectancy of 50-year-old white women treated with long-term hormone replacement (Grady et al. 1992)

Variable	Life expectancy (years)	Net change in life expectancy (years)		
		Estrogen only	Estrogen+ progestin[a]	Estrogen+ progestin[b]
No risk factors	82.8	+0.9	+1.0	+0.1
With hysterectomy	82.8	+1.1		
With history of coronary heart disease	76.0	+2.1	+2.2	+0.9
At risk for coronary heart disease	79.6	+1.5	+1.6	+0.6
At risk for breast cancer	82.3	+0.7	+0.8	–0.5
At risk for hip fracture	82.4	+1.0	+1.1	+0.2

[a]Assuming that the addition of a progestin to the estrogen regimen does not alter any of the risks for disease seen with estrogen therapy, except to prevent the increased risk due to endometrial cancer (relative risk for endometrial cancer estimated to be 1.0).
[b]Assuming that the addition of a progestin to the estrogen regimen provides only two-thirds of the coronary heart disease risk reduction afforded by estrogen therapy (relative risk for coronary heart disease).

The effect that HRT, which generally includes progestogen to prevent endometrial cancer, has on lipids is inconclusive (Table 4).

Estrogen administered to decrease TC and increase HDL-C can effect an improvement in hyperlipidemia within 2 months. However, the recently developed 3-hydroxy-3-methylglutaryl coenzyme A reductase inhibitors [pravastatin sodium (Mevalotin, Sankyo), etc.] (Honjo et al., 1992) and other antihyperlipidemic drugs can also improve hyperlipidemia within 2 weeks. In critical cases where TC exceeds 260 mg/dl, antihyperlipidemic drugs should be selected as the first choice, or should be used concomitantly with estrogen. In the case of high TG levels, large doses of estrogen should be given with caution. In such cases, bezafibrate (Bezatol SR, Kissei) is recommended.

For mild to medium hyperlipidemia, especially if accompanied by high TC, vasomotor disorders including hot flushes, or osteoporosis, low-dose estrogen such as conjugated estrogen (Premarin, Ayerst) is recommended at a dose of 0.625 mg/day; alternatively, an estrogen

Table 4. Changes in lipid profiles in women receiving combined continuous estrogen and progestin replacement therapy (adapted from Udoff et al. 1995)

Primary author (year)	Estrogen	Progestin	Average duration of treatment (months)	TC	LDL-C	HDL-C
Weinstein	CEE (0.625 mg)	MPA (2.5 mg)	3	↓	↓	↓
(1987)	CEE (0.625 mg)	MPA (5.0 mg)	3	↔	↔	↔
Weinstein	CEE (0.625 mg)	MPA (2.5 mg)	12	↔	↔	↔
(1990)	CEE (0.625 mg)	MPA (5.0 mg)	12	↔	↔	↔
Prough (1987)	CEE (0.625 mg)	MPA (2.5 mg)	9	↔	↔	↔
Kable (1990)	CEE (0.625 mg)	MPA (2.5–5.0 mg)	3	↓	↓	↑
Luciano	CEE (0.625 mg)	MPA (2.5 mg)	13	↓	↓	↔
(1993)1	CEE (0.625 mg)	MPA (5.0 mg)	12	↓	↓	↔
Lobo (1994)	CEE (0.625 mg)	MPA (2.5 mg)	13	↓	↓	↑
	CEE (0.625 mg)	MPA (5.0 mg)	13	↓	↓	↑
PEPI (1995)	CEE (0.625 mg)	MPA (2.5 mg)	36	↓	↓	↑
Luciano (1988)	CEE (0.625–1.25 mg)	MPA (10 mg)	3	↓	↔	↔
Clisham	CEE (0.625 mg)	MPA (10 mg)	3	↔	↔	↔
(1991)	CEE (0.125 mg)	MPA (10 mg)	3	↔	↔	↔
MacLennan	CEE (0.3–0.625 mg)	MPA (2.5 mg)	12	↔	↔	↔
(1993)	CEE (0.3–0.625 mg)	Levonorgestrel (30 μg)	12	↔	↔	↔
	CEE (0.3–0.625 mg)	NEA (0.35 mg)	12	↔	↔	↔
Staland (1985)	E_2 (2 mg)+E3 (1 mg)	NEA (1 mg)	18	↓		
Mattson (1985)	E_2 (2 mg)+E3 (1 mg)	NEA (1 mg)	12	↓	↓	↓
Farish (1987)	E_2 (2 mg)	NEA (1 mg)	48	↔	↔	↔
	E_2 (2 mg)+E_3 (1 mg)	NEA (1 mg)	48	↔	↔	↔
Jensen (1987)	E2 (2 mg)	NEA (1 mg)	12	↓	↓	↔
Christiansen (1990)	E_2 (2 mg)	NEA (1 mg)	60	↓	↓	↔
Christiansen (1990)	E_2 (2 mg)	NEA (1 mg)	12	↓	↓	↓
Keller (1992)	E_2 (0.05 mg)	NEA (0.25 mg)	6	↔	↓	↔
Munk-Jensen (1994)	E_2 (2 mg)	NEA (1 mg)	24	↓	↓	↓
Williams (1990)	EE_2 (0.5–20 μg)	NEA (05–1.0 mg)	12	↔	↔	↔
Hargrove (1989)	E_2 (0.7–1.05 mg)	Progesterone (200–300 mg)	12	↓		↑
Marslew (1991)	E_2 val (2 mg)	CPA (1 mg)	24	↓	↓	↔
Metka (1992)	E_2 val (2 mg)	CPA (1 mg)	6	↔	↔	↔
Cano (1991)	E_2 val (2 mg)	MPA (2.5 mg)	8	↓	↓	↑
March (1994)	E_2 (1 mg)	Desogestrel (0.15 mg)	12	↓	↓	↓
Voetberg	E_2 (2 mg)	DDG (2.5 mg)	6	↓	↓	↑
(1994)	E_2 (2 mg)	DDG (5.0 mg)	6	↓	↓	↑
	E_2 (2 mg)	DDG (10 mg)	6	↓	↓	↑
	E_2 (2 mg)	DDG (15 mg)	6	↓	↓	↔

CEE, conjugated estrogen; MPA, medroxyprogesterone acetate; E_3, estriol; NEA, norethindrone acetate; E_2, estradiol; EE_2, ethinyl estradiol; val, valerate; CPA, cyproterone acetate; DDG, dydrogesterone.
Arrows represent statistically significant differences between pre- and posttreatment values: ↑, increase; ↓, decrease; ↔, no change.
[a]For reference details, see Udoff et al. 1995.

patch (Estraderm TTS, Ciba-Geigy; 2 mg) is also effective, at one patch per administration.

6.6 Conclusion

In the United States and in European countries, HRT is very popular, and treatment of cardiovascular diseases by HRT is also included. A few internists (specialists of cardiovascular disease and the others) have now started applying HRT for the treatment of cardiovascular diseases in Japan.

As the world is faced with an increasingly aging population, the use of HRT in enhancing the quality of life of elderly women deserves further study, which may produce even wider applications for its use (Honjo 1995, Honjo 1998).

References

Grady D, Rubin SM, Petitti DB, Fox CS, Black D, Ettinger B, Ernster VL, Cummings SR (1992) Hormone therapy to prevent disease and prolong life in postmenopausal women. Ann Intern Med 117:1016–1037

Hirota K, Honjo H, Shintani M (1995) Factors on menopause – a study in endocrine research group (1994). Adv Obstet Gynecol 47:389–392

Honjo H (1991) Clinical results of mevalotin: study of usefulness against hypercholesterolemia in the climacteric woman. J Am Med Assoc (Japanese version) 12 (appendix):26–29

Honjo H (1995) Climacteric and elderly out-patients' manual – invitation to queen's corner, 1st edn, 2nd issue. Kinpodo, Kyoto

Honjo H (1998) More information about HRT, Chizin-sha, Kyoto

Honjo H, Tanaka K, Urabe M, Naitoh K, Ogino Y, Yamamoto T, Okada H (1992) Menopause and hyperlipidemia: pravastatin lowers lipid levels without decreasing endogenous estrogens. Clin Ther 14:699–707

Honjo H, Tanaka K, Kashiwagi T, Okada H, Takanashi K, Yoshizawa I (1994) Women of menopausal age and hyperlipidemia: effects of estrogens and an HMG-CoA reductase inhibitor. In: Yamamoto A (ed) Multiple risk factors in cardiovascular disease. Churchill Livingstone, Tokyo, pp 81–84

Honjo H, Tanaka K, Kashiwagi T, Urabe M, Okada H, Hayashi M, Hayashi K (1995) Senile dementia Alzheimer's type and estrogen. Horm Metab Res 27:204–207

Honjo H, Urabe M, Iwasa K,Okubo I, Tsuchiya H, Kikuchi N, Yamamoto T, Fushiki S, Mizuno T, Nakajima K, Hayashi M, Hayashi K (1997) Estrogen treatment for senile dementia – Alzheimer's type. In: Wren BG (ed) Progress in the management of the menopause. Parthenon, New York, pp 302–307

Honjo H, Iwasa K, Urabe M (1998) Clinical studies of estrogen therapy for dementia, J Brit Menopause Soc 4:12–17

Japanese Ministry of Health and Welfare, 31 August 1996; Population Development Statistics for 1995, 404–405

Japanese Ministry of Health and Welfare, 28 August 1998; Simple life table in 1997

Koseisho (Ministry of Health and Welfare) (Jan 1993). Outline of the 4th Cardiac Disease Basic Survey (Nov 1990) by the Disease Control Division, Health Service Bureau, Mainichi (newspaper), 4 Feb 1993

Mahley RW (1988) Apolipoprotein E; cholesterol transport protein with expanding role in cell biology. Science 240:622–629

Ohkura T, Teshima Y, Isse K, Matsuda H, Inoue T, Sakai Y, Iwasaki N, Yaoi Y (1995) Estrogen increases cerebral and cerebellar blood flows in postmenopausal women. Menopause 2:13–8

Paganini-Hill A, Henderson VW (1996) Estrogen replacement therapy and risk of Alzheimer's disease. Arch Intern Med 156:2213–2217

Rosano GMC, Sarrel PM, Poole-Wilson PA, Collins P (1993) Beneficial effect of estrogen on exercise-induced myocardial ischaemia in women with coronary artery disease. Lancet 342:133–6

Tang M-X, Jacobs D, Stern Y, Marder K, Schofield P, Gurland B, Andrews H, Mayeux R (1996) Effect of oestrogen during menopause on risk and age at onset of Alzheimer's disease. Lancet 348:429–432

Udoff L, Langenberg P, Adashi EY (1995) Combined continuous hormone replacement therapy: a critical review. Obstet Gynecol 86:306–16

Urabe M, Yamamoto T, Kashiwagi T, Okubo T, Tsuchiya H, Iwasa K, Kikuchi N, Yokota K, Hosokawa K, Honjo H (1996) Effect of estrogen replacement therapy on hepatic triglyceride lipase, lipoprotein lipase and lipids including apolipoprotein E in climacteric and elderly women. Endocr J 43:737–742

7 HRT and the Vascular System – Direct Action of Estrogen on Vascular Smooth Muscle Cells

M. Nozaki

7.1 Introduction

It is well known that cardiovascular disease is the leading cause of morbidity and mortality in postmenopausal women (Bush et al. 1987; Henderson et al. 1988). It is also generally accepted that treatment with estrogen reduces the mortality rate due to cerebro- and cardiovascular diseases (Stampfer et al. 1985; Stampfer et al. 1991; Paganini-Hill et al. 1988; Sarrel et al. 1994; Williams et al. 1990; Knopp 1988). From this point of view, it is important to clarify the mechanism of alteration in the vascular system that lacks estrogen, when dealing with postmenopausal women.

The aim of this study is to elucidate whether or not estrogen influences the structure and function of vascular tissue, especially vascular smooth muscle (VSM) cells. In the study of vascular structure, the effects of estradiol on the proliferation of synthetic types of VSM cells, excised from rabbit thoracic aorta, were examined. In the study of vascular function, the effects of estradiol on the contraction–relaxation mechanism of rabbit basilar VSM cells were investigated.

7.2 The Effects of Estradiol on the Vascular Structure

There are two potentially different phenotypic states of the VSM cell, the contractile and the synthetic phenotype. Contractile VSM cells are found in normal vascular tissue, are highly differentiated, and contain many contractile fibers. On the other hand, synthetic VSM cells are not usually found in normal vascular tissue but are found in atherosclerotic lesions (Ross 1993). Their differentiation is low, they have little contractility, but they have a higher level of proliferation.

As shown in Fig. 1, besides various factors, such as injuries to the endothelium, growth factors, and chemoattractants from macrophages and endothelium, the conversion of SMC is one of the most important phenomena in the development of atherosclerosis, where the contractile phenotype of SMC converts to the synthetic phenotype of SMC. After conversion of contractile SMC to synthetic SMC, transmigration of SMC to the subendothelial space is found in atherosclerotic lesions (Casscells 1992).

By electron microscope, no differences could be seen in contractile VSM cells and collagen fibers of rabbit basilar arteries in either the 12-week estrogen-lack or -replacement group (Nozaki 1997). The contractile VSM cells without remarkable morphological change were used in experiments on contraction.

The rabbit thoracic aorta was used in the experiment on the proliferation of the synthetic type of VSM cells. The thoracic aorta was excised, and SMC was dissociated in collagenase, and then a primary culture was made. After passage, SMC was confirmed by immunofluorescent stain with α-SMA. The inhibitory effect of 17β-estradiol on thymidine uptake in the proliferation of synthetic-type VSM cells was studied in cultures of VSM cells. After the addition of 17β-estradiol and thymidine, DNA synthesis was determined with a liquid scintillation counter. The thymidine uptake was inhibited by incubation with 17β-estradiol in a concentration-dependent manner (Nozaki 1997).

The patch-clamp method was used to test whether estrogen has an inhibitory effect on characteristic calcium (Ca) currents recorded from

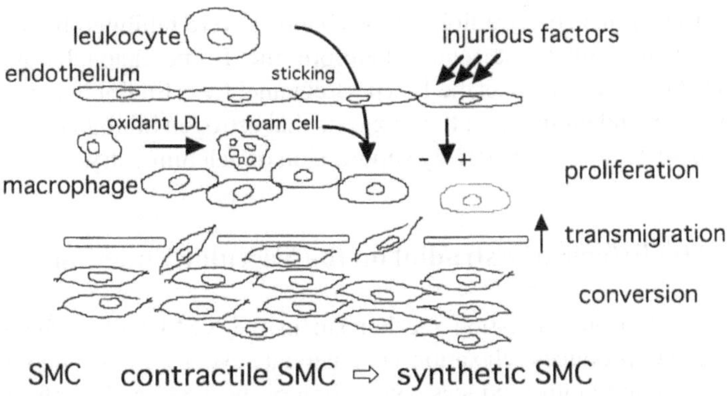

Fig. 1. The development of atherosclerosis. Besides various factors, such as injuries to the endothelium, growth factors, and chemoattractants from macrophages and endothelium, the conversion of SMC is one of the most important phenomena in the development of atherosclerosis; the contractile phenotype of SMC is converted to the synthetic phenotype of SMC. After conversion of contractile SMC to synthetic SMC, transmigration of SMC to the subendothelial space takes place in atherosclerotic lesions

cultured VSM cells (Nozaki 1997). With the whole-cell clamp method, inward calcium currents were measured through a pipette attached to the cell membrane.

When the patch-clamp method was used in both freshly isolated and cultured VSM cells from rabbit thoracic aorta, distinct wave forms were recorded for three types of calcium currents: a high-voltage-activated current with slow inactivation (L-type 1) in freshly isolated cells, a high-voltage-activated current with fast inactivation (L-type 2), and a low-voltage-activated current (T type), characteristically appearing in cultured synthetic VSM cells. In the freshly isolated VSM cells, L-type-1 channels make up 90%.

On the other hand, in cultured VSM cells, L-type-2 and T-type channels are recorded in addition to L-type-1 channels. It has been reported that calcium entry is required in the cell proliferation. L-type-2 and T-type calcium channels are, therefore, the characteristic channels in the proliferation of VSM cells. When the VSM cells were incubated with 10 μM of 17β-estradiol, the amplitude of the current in the L-type channels was inhibited by 20%. When the VSM cells were incubated with 10 μM of 17β-estradiol, T-type channels were inhibited by 40% in the amplitude of the currents. Therefore, the T-type channel was suppressed twice as much as the L-type-1 channel by estradiol in a concentration-dependent manner; this suggests that estrogen inhibits the proliferation of VSM cells partly by suppression of calcium entry.

7.3 The Effects of Estradiol on the Vascular Function

The contraction–relaxation mechanism in the VSM cell is shown in Fig. 2. NO produces relaxation via guanylate cyclase in the endothelial cell, and endothelin produces contraction of the VSM cell. In the VSM cell, various receptors and channels in the membrane, as well as at the SR membrane, regulate the calcium concentration, resulting in contraction or relaxation. The basilar artery of the same rabbit was used in the following experiments to compare the contractile properties of the VSM cells in the ovariectomized rabbit and in the estrogen-replaced one, and to investigate the mechanism of the acute action of estrogen on the contractile properties of VSM cells excised from the OVX rabbit, in addition to that of its chronic action. The isometric-tension recording method was used and the ring preparations of the artery were placed in a bath solution (Nozaki and Ito 1986; Nozaki and Ito 1987).

Acetylcholine inhibited the high-potassium contraction in the endothelium-intact vessel. On the other hand, acetylcholine enhanced high-potassium contraction in the endothelium-denuded vessel. From these facts, we can see that acetylcholine acts on VSM cells as an endothelium-dependent relaxant by the NO-generating pathway. No morphological changes were detected by electron microscope in the basilar artery of the OVX rabbit after a 12-week estrogen deficiency in the VSM cells and collagen fibers. On the other hand, the contractile property of VSM cells in the OVX rabbit was changed compared to that of the estrogen-replaced one. When acetylcholine relaxed arterial tissue precontracted by excess potassium solution in a concentration-dependent manner, a 12-week estradiol treatment enhanced the endothelium-dependent relaxation produced by acetylcholine. Furthermore, estradiol treatment enhanced the relaxation produced by an NO donor such as NOR1 in endothelium-removed arterial tissue. The contractile property of VSM cells in the OVX rabbit was changed compared to that of the

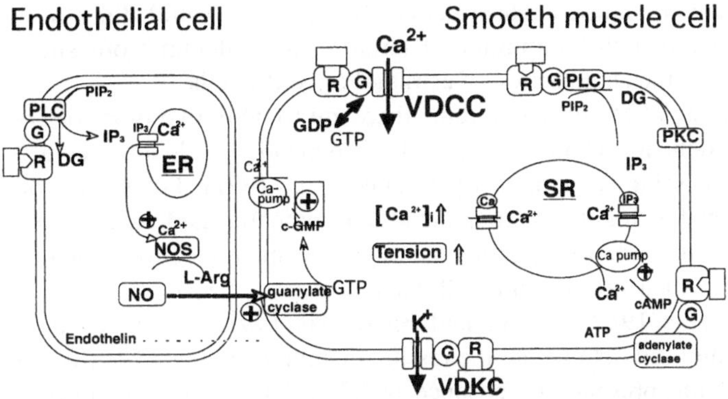

Fig. 2. The contraction–relaxation mechanism in the VSM cell. In the endothelial cell, NO produces relaxation via guanylate cyclase and endothelin produces contraction of the VSM cell. In the VSM cell, various receptors and channels in the membrane, as well as at the SR membrane, regulate the calcium concentration; this results in contraction or relaxation

estrogen-replaced one. When an NO donor relaxed arterial tissue precontracted by excess potassium solution in a concentration-dependent manner, the 12-week estradiol-treatment enhanced endothelium-independent relaxation produced by the NO donor. These results, concerning the chronic (genomic) action of estrogen, suggest that estrogen enhances not only endothelium-dependent relaxation but also the sensitivity towards NO donors in VSM cells.

In the experiment on the acute (nongenomic) action of estrogen, estradiol quickly relaxed arterial tissue precontracted by excess potassium solution in a concentration-dependent manner (Ogata et al. 1996). The amplitude of contraction recovered after washout. Estradiol relaxed arterial tissue precontracted by excess potassium solution in a concentration-dependent manner in the experiment on acute action. However, there was no difference in the inhibition of contraction in the presence or absence of endothelium. Therefore, endothelium is not responsible for the acute action of estradiol on the relaxation of VSM cells.

In calcium-free solution, caffeine induces calcium release from SR, the calcium store site of SMC, to increase intracellular calcium concentration, resulting in contraction of VSM cells. In calcium-free solution, caffeine produced a phasic contraction, but estradiol did not significantly affect the amplitude. Therefore, the calcium store site is not linked to the acute action of estradiol on VSM cells.

In the VSM cell membrane, there are roughly two types of voltage-dependent ion channels, the voltage-dependent calcium channels and the potassium channels (Sperelakis et al. 1991). Potassium ions are responsible for the resting membrane potential and calcium ions are responsible for the action potential. The microelectrode method was used to measure the intracellular potential of VSM cells (Nozaki and Sperelakis 1989; Nozaki and Sperelakis 1991; Nozaki et al. 1990). Estradiol did not change the resting membrane potential of the artery in either the presence or absence of TEA. TEA is the inhibitor of the potassium channel; it depolarizes the membrane. Action potentials observed in the presence of TEA were abolished by estradiol.

These results show that the acute action, the nongenomic action, of estradiol on the VSM cells may cause the inhibition of voltage-dependent calcium channels. Because these types of calcium channels are reported to be coupled with G proteins, GDPβS (guanosine 5'-diphosphate βS) and GTPγS (guanosine 5'-triphosphate γS) were used in the following experiments. The patch-clamp method was used to test the effects of estradiol on voltage-dependent calcium channels and calcium-inward currents (Ohya et al. 1990). Estradiol inhibited the voltage-dependent calcium current. The amplitude of the calcium current was restored after washout. In the following experiment, estradiol inhibited the voltage-dependent calcium current in a concentration-dependent manner. Estradiol did not inhibit calcium channels after the addition of the nonsoluble agonist, GDPβS, to the pipette.

On the other hand, when GTPγS, the stimulatory substance of the G protein, was placed in the pipette, the inhibitory action of estradiol on the calcium current was enhanced in a concentration-dependent manner. GTPγS therefore enhanced the inhibitory effects of estradiol on the calcium channels, whereas GDPβS canceled it. In the final experiment, pertussis toxin (PTX), in the pipette, totally prevented the inhibitory action of estradiol on the calcium current. These results strongly suggest that estradiol relaxes arterial tissue by the inhibition of voltage-dependent calcium channels via a PTX-sensitive GTP-binding protein in the VSM cell membrane.

7.4 Conclusion

In conclusion, the functional changes due to estrogen depletion appeared earlier than the morphological changes in the VSM cells. Estrogen suppressed the proliferation of the synthetic type of VSM cells, in part by the inhibition of calcium entry through the characteristic calcium channel. Estrogen has a genomic action on VSM cells; it acts as an enhancer of endothelium-dependent and -independent relaxation via the NO-generating and -receiving pathway. Estrogen also has nongenomic action on vascular tissue; it acts as a vasodilator through the inhibition of calcium channels via a PTX-sensitive G protein in the VSM cell membrane.

Among various actions, such as improvement of lipid profile, increasing prostacyclin, inhibition of LDL oxidation, decreasing elastin and collagen in the aortic wall, estrogen also inhibits the proliferation of SMC, to suppress the development of atherosclerosis (Fig. 3). On the other hand, estrogen acts as a vasodilator, mainly on SMC, to reduce vascular tonus. Estrogen also acts as an enhancer of relaxation, both on SMC and endothelium, to reduce vascular tonus (Fig. 4). Among various vasodilating and vasoconstricting substances, estradiol plays a key role in the regulation of vascular tonus. Angiotensin II, thromboxane A_2, endothelin, PGH_2 and superoxide increase vascular tonus (Fig. 5). On the other hand, NO, prostacyclin, CNP, EDIHF, and estradiol decrease vascular tonus. Almost all vasoconstricting substances and the other growth factors stimulate the proliferation of VSM cells (Fig. 6). And almost all vasodilating substances and estradiol inhibit the proliferation of VSM cells. Estradiol also plays a key role in the regulation of VSM cell proliferation, to protect against arteriosclerosis besides having its own action on the lipid metabolism.

Menopause may lead to vasoconstriction in vascular tonus, followed by VSM cell proliferation towards atherosclerosis. The increased vascular tonus associated with menopause should be corrected by estrogen replacement. HRT sets this balance right in postmenopausal women. HRT should be considered even in normo-lipidemic postmenopausal women, because functional changes may have already occurred in the vascular tissue without remarkable, morphological changes.

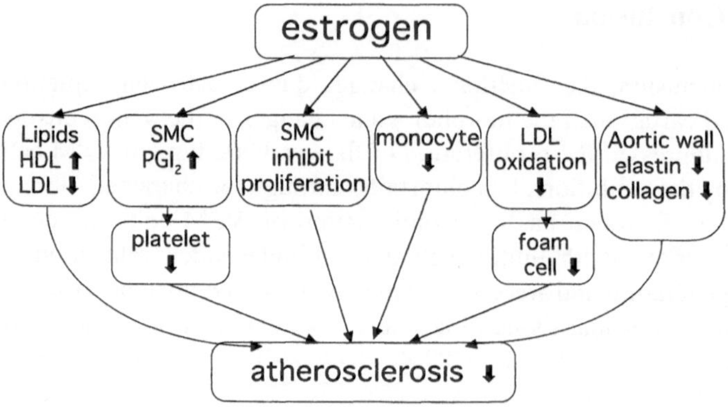

Fig. 3. The antiatherosclerotic effects of estrogen

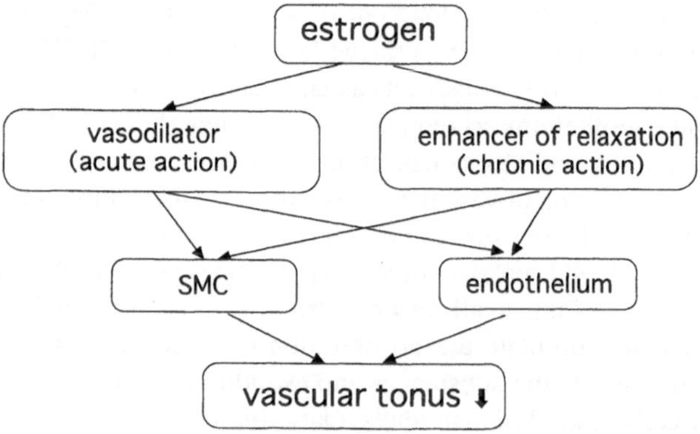

Fig. 4. The effect of estrogen on vascular tonus

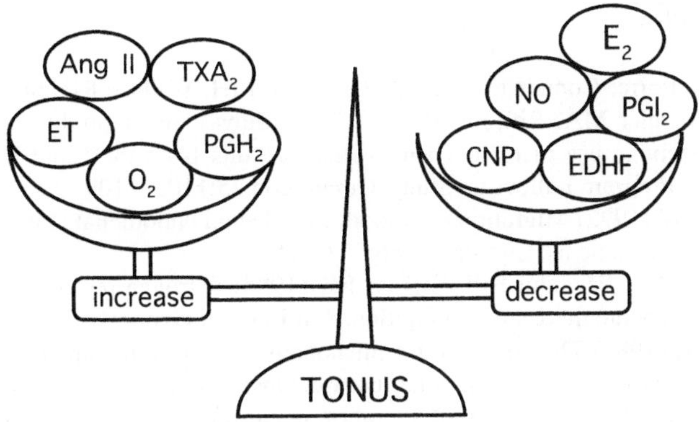

Fig. 5. The role of estradiol in vascular tonus

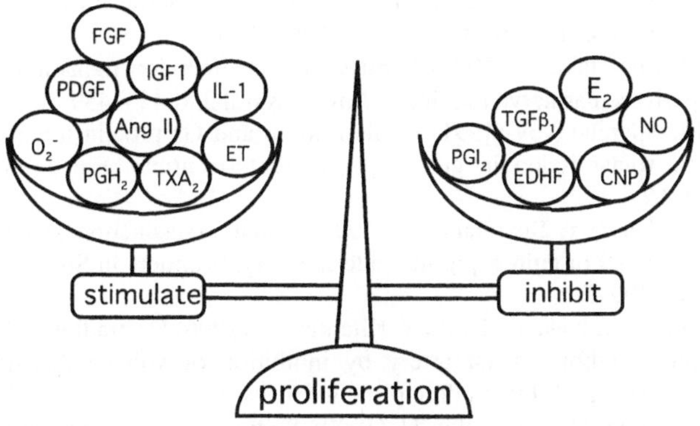

Fig. 6. The role of estradiol in VSMC proliferation

References

Bush TL, Barret-Connor E, Cowan LD, Criqui MH, Wallace RB, Suchindran CM, Tyroler MA, Rifkind BM (1987) Cardiovascular mortality and noncontraceptive use of estrogen in women: results from the Lipid Research Clinics Program Follow-up Study. Circulation 75:1102–1109

Casscells W (1992) Migration of smooth muscle and endothelial cells. Critical events in restenosis. Circulation 86:723–729

Henderson BE, Paganini-Hill A, Ross RK (1988) Estrogen replacement therapy from acute myocardial infarction. Am J Obstet Gynecol 159:312–317

Knopp RH (1988) The effects of postmenopausal estrogen therapy on the incidence of arteriosclerotic vascular disease. Obstet Gynecol 72:23S–30S

Nozaki M (1997) Mechanism of alteration in cardiovascular system by menopause. Acta Obstet Gynaec Jpn 552–562

Nozaki M, Ito Y (1986) Menstrual cycle and sensitivity of human fallopian tube to prostaglandins. Am J Physiol 251:1126–1136

Nozaki M, Ito Y (1987) Changes in physiological properties of rabbit oviduct by ovarian steroids. Am J Physiol 252:1059–1065

Nozaki M, Sperelakis N (1989) Pertussis toxin effects on transmitter release from perivascular nerve terminals. Am J Physiol 256:455–459

Nozaki M, Sperelakis N (1991) Cholera toxin and Gs protein modulation of synaptic transmission in guinea pig mesenteric artery. Eur J Pharmacol 197:57–62

Nozaki M, Ohya Y, Sperelakis N (1990) Signal transduction at adrenergic nerve terminal in guinea-pig mesenteric artery. Frontiers in Smooth Muscle Research 499–506

Ogata R, Inoue Y, Nakano, H, Ito Y, Kitamura K (1996) Oestradiol–induced relaxation of rabbit basilar artery by inhibition of voltage–dependent Ca channels through GTP–binding protein. British J Pharmacol : 117;351–359

Ohya Y, Nozaki M, Sperelakis N (1990) Sodium and calcium currents recorded from single uterine smooth muscle cells freshly isolated from pregnant rat–Frontiers in Smooth Muscle Research 659–663

Paganini–Hill A, Ross RK, Henderson BE (1988) Postmenopausal oestrogen treatment and stroke: a prospective study. Br Med J 295:519–522

Ross R (1993) The pathogenesis of atherosclerosis: a perspective for the 1990s. 362:801–809

Sarrel PM, Lufkin EG, Ousler MJ, Keef D (1994) Estrogen actions in arteries, bone and brain. Sci Am Sci Med 1:44–53

Sperelakis N, Inoue Y, Nozaki M, Ishikawa S (1991) Neuromuscular transmission at adrenergic nerve terminals with vascular smooth muscle in guinea-pig mesenteric artery. In: Sperelakis N, Kuriyama H (eds) Ion channels of

vascular smooth muscle cells and endothelial cells. Elsevier, New York, pp 3–15
Stampfer MJ, Willett WC, Colditz GA, Rosner B, Speizer FE, Hennekens CH (1985) A prospective study of postmenopausal estrogen therapy and coronary heart disease. N Engl J Med 313:1044–1049
Stampfer MJ, Colditz GA, Willett WC, Manson JE, Rosner B, Speizer FE, Hennekens CH (1991) Postmenopausal estrogen therapy and cardiovascular disease. N Engl J Med 325:756–762
Williams JK, Adams MR, Klopfenstein HS (1990) Estrogen modulates responses of atherosclerotic coronary arteries. Circulation 81:1680–1687

8 The Place of Estriol Add-Back Therapy in Women Treated with GnRH Agonist, with Reference to Bone Protection

T. Yano, A. Kikuchi, H. Matsumi, Y. Wang, H. Nakayama, and Y. Taketani

8.1 Introduction

Chronic administration of gonadotropin-releasing hormone agonist (GnRHa) leads to inhibition of the pituitary–gonadal axis, resulting in a marked suppression of ovarian estrogen production (Schally 1989). Accordingly, GnRHa has been used widely in the treatment of a variety of estrogen-dependent disorders such as endometriosis (Wheeler et al. 1992) and uterine leiomyoma (Adamson 1992). However, long-term GnRHa treatment is not recommended because of concern about bone loss due to hypoestrogenism (Adashi 1994; Surrey 1995). On the other hand, long-term GnRHa treatment is indispensable for women with serious medical illnesses (e.g., leukemia, aplastic anemia, and severe cardiac disease) that would contraindicate surgery. Women approaching menopause are also candidates for long-term treatment. It has recently been proposed that, in order to minimize bone loss caused by hypoestrogenism, but without undermining the therapeutic efficacy of GnRHa, low doses of estrogen and/or progestin should be added back, thus making long-term GnRHa treatment possible; this is known as "add-back therapy" (Adashi 1994; Surrey 1995). To date, however, there is no agreement on an optimal GnRHa/steroid add-back regimen.

Estriol (E3) is a short-acting estrogen because it is rapidly conjugated in the liver (Schiff et al. 1980) and the duration of nuclear-receptor binding is relatively short (Clark et al. 1977). A recent study has demonstrated that the binding affinity of E3 for ERα and ERβ is remarkably lower than that of estrone (E1) or estradiol (E2) (Kuiper et al. 1997). The duration of plasma E3 elevation is 3–4 hours after an oral administration (Englund et al. 1982). Estrogens with relatively low binding affinity or with short-duration receptor binding may be the most desirable add-back regimen for women administered GnRHa. E3 would, therefore, be expected to be an appropriate estrogen used for add-back therapy. Notably, E3 exerts either estrogenically antagonistic or agonistic effects (Clark et al. 1977; Heimer 1987; Melamed et al. 1997). When given alone, it acts as an agonist, the potency of which depends on the dosage size and the frequency of administration (Heimer 1987). When given in conjunction with another more potent estrogen such as E2, it works as an antagonist (Melamed et al. 1997). E3 is thought to be safer than E1 and E2, especially in that it causes little, if any, change in uterine tissues such as endometrial proliferation and hyperplasia (Tzingounis et al. 1978; Grasso et al. 1982; Minaguchi et al. 1996). However, efficacy of E3 on preventing bone loss remains to be established.

In the animal experiments, in search among various natural estrogens for an optimal estrogen for add-back therapy, we compared the effects of E1, E2, and E3 in bone-protective and uterine effects in rats administered a long-acting GnRHa. This study particularly centered on cancellous bone histomorphometry in proximal tibiae and serum markers for bone metabolism, to get insight into the cellular, architectural, and metabolic changes in the bone, in addition to the measurement of bone mineral density (BMD) of lumbar vertebrae (L1–L5) and femoral bone. In the clinical study, we investigated the effects of E3 in combination with GnRHa on uterine leiomyomas, with special reference to BMD of lumbar spine and endometrial thickness.

8.2 Materials and Methods

8.2.1 Animal Experiment

8.2.1.1 Chemicals

Microcapsules of GnRHa leuprorelin acetate, which provide continuous drug release for 4 weeks after injection, were supplied by Takeda Pharmaceutical Co. (Osaka, Japan). Pellets of E1, E2, and E3 (0.5 mg, 60-day release) were purchased from Innovative Research of America (Sarasota, Fla., USA). All other chemicals, unless otherwise mentioned, were obtained from Sigma Chemical Co. (St. Louis, Mo., USA).

8.2.1.2 Animals

Three-month-old female Sprague–Dawley rats, weighing 250 g, were obtained from Takasugi Experimental Animal Inc. (Saitama, Japan). The rats were housed in hanging wire cages, maintained at 25 °C with a 12 h light/12 h dark schedule, had continuous access to food (MF Diet: 1.15% calcium and 0.88% phosphorus, Oriental Yeast Co., Kanagawa, Japan), and had water ad libitum. The rats were divided into 5 groups (6 animals per group) and received the following treatments: group 1, the control: 0.2 ml injection vehicle alone was subcutaneously administered every 4 weeks; group 2: leuprorelin acetate microcapsules suspended in 0.2 ml injection vehicle were subcutaneously administered at a dose of 1 mg/kg body weight, every 4 weeks; and groups 3, 4, and 5 had, respectively, in addition to GnRHa administration, E1, E2, or E3 pellets subcutaneously implanted, on the backs of animals, with a trocar at the time of the third GnRHa injection. The estrogen pellets were designed to release about 8 μg/day in rats weighing 250 g; this corresponds to a clinical therapeutic dose (approx. 1–2 mg/day) of natural estrogen in women. The treatment was continued for 16 weeks and GnRHa was injected 4 times in groups 2–5.

At the end of the treatment, the rats were weighed and sacrificed by decapitation under ether anesthesia, and trunk blood was collected. All the blood samples were centrifuged and serum was stored at –80°C until assayed. Uteri were quickly removed, cleaned of surrounding tissue, and weighed. Lumbar vertebrae (L1–L5) and the left femoral bone were removed for the BMD measurement. Right proximal tibiae were removed, dissected free of musculature, and fixed in 70% ethanol at room temperature for bone histomorphometry. Serum levels of osteocalcin, E1, E2, and E3 were determined by radioimmunoassay (RIA). Serum alkaline phosphatase (ALP) activities were determined by a standard colorimetric method. BMD was measured by dual-energy X-ray absorptiometry (DXA) with DPX-L (Lunar Co. Madison, WI, USA). The lateral lumber vertebrae (L1–L5) and the antero-posterior left femoral bone were scanned to determine their BMD.

8.2.1.3 Bone Histomorphometric Analysis

Tetracycline (Wako Pure Chemicals, Osaka, Japan) was subcutaneously injected into each rat at a dose of 25 mg/kg body weight on the 14th day before sacrifice, and calcein (Wako Pure Chemicals, Osaka, Japan) was subcutaneously injected into each rat at a dose of 13 mg/kg body weight on the 7th day before sacrifice. After one month of fixation in 70% ethanol, the right tibia specimens were stained with Villanueva bone stain solution for 10 days. They were dehydrated in a series of increasing concentrations of ethanol, defatted in acetone, and embedded in methyl methacrylate (MMA, Wako Pure Chemicals) mixture. Longitudinal sections of 3 μm were made with RM2065 microtome (Leica Instruments GmbH, Nussloch, Germany) and mounted on poly(L-lysine)-coated glass slides. The sampling site was situated in the secondary spongiosa of the metaphyseal region of the proximal tibia at distances greater than 1 mm from the growth plate–metaphyseal junction, to exclude the primary spongiosa. A total metaphyseal area of 0.5 mm^2 was sampled for each section. Photomicrographs were taken of each sampled metaphyseal area under a microscope (Nikon, Tokyo, Japan). Histomorphometric measurements were performed with a digitizer, Oscon A4-40 (Pretec, Tokyo, Japan), interfaced with a PC 9801 computer (NEC, Tokyo, Japan). Cancellous bone areas and surface lengths in the photomicrographs were traced with a cursor on a digitizing tablet and calculations were done by the computer. Bone histomorphometric parameters were measured according to the report of the ASBMR histomorphometry nomenclature committee (Parfitt et al. 1987).

8.2.2 Clinical Study

8.2.2.1 Subjects

Twelve premenopausal women with uterine leiomyomas and ovulatory cycles were enrolled in this study. The mean age (±SEM) of the subjects at the beginning of the treatment was 47.4±0.6 years (range 44–51) and the body mass index (BMI) was 23.1±0.7 (kg/m^2). None of the subjects had any diseases or medications known to affect bone metabolism or renal function. The subjects received a GnRHa depot, leuprorelin acetate depot, at a dose of 3.75 mg s.c. every month for 6 months. The initial dose of GnRHa was given on day 2–5 of the menstrual cycle. After the first 2 months of treatment, the subjects were allocated randomly to 2 groups and received either the GnRHa depot alone (non-add-back group; n=6) or the GnRHa depot monthly plus 4 mg oral E3 (Mochida Pharmaceutical Co., Tokyo, Japan) daily (add-back group; n=6) for the remaining 4 months. BMD of the lumbar spine (L2–L4) was measured by DXA. As markers of bone formation, serum osteocalcin and bone-specific alkaline phosphatase (B-ALP) levels were measured. Urinary CrossLaps and deoxypyridinoline were measured to evaluate the status of bone resorption. Serum and daytime urine specimens were collected in the morning on the day of every GnRHa administration and 1 month after the last administration. Serum levels of E2 and osteocalcin were determined by RIA. Serum B-ALP levels were determined by the polyacrylamide gel disk electrophoresis technique (AlkPhor System; Quantimetrix Co., Redondo Beach, Calif., USA). Urinary CrossLaps levels were measured with an ELISA kit (Osteometer Co., Copenhagen, Denmark). Urinary levels of deoxypyridinoline were measured by high-pressure liquid chromatography (HPLC). All urinary parameters were standardized by urinary creatinine (Cr) concentration measured by a standard colorimetric method.

8.2.2.2 Leiomyoma Volume and Endometrial Thickness

Volume of uterine leiomyomas and endometrial thickness were measured by transvaginal ultrasound (with a 5 MHz probe) every month. Before the treatment, the size and diagnosis of leiomyoma were delineated by magnetic resonance imaging. The volume of each leiomyoma nodule was calculated by the formula used for measuring the volume of ellipsoid tumors:

$V=(1/6)\ \pi \times D \times D1 \times D2 \times D3$,

where D1, D2 and D3 are the three largest diameters. The value of the largest nodule in each patient was used for statistical analysis.

8.3 Results

8.3.1 Animal Experiment

Rats given GnRHa alone exhibited a marked reduction in uterine weight, by 81.8% compared with the control rats. These changes were eliminated by the concomitant treatment with E1 or E2. In contrast, E3 did not prevent GnRHa-induced uterine atrophy. In the control rats, serum E2 levels range from 10.0 pg/ml to 20.2 pg/ml. In GnRHa-treated rats, serum E2 levels fell to undetectable levels. As expected, considerable levels of serum E2 and E3 were detected in rats treated with E2 and E3, respectively. E1-treated rats exhibited easily detectable amounts of serum E1 and E2.

Serum levels of ALP and osteocalcin in GnRHa-treated rats were significantly higher than those in the control rats. These changes by GnRHa were abrogated by concomitant administration of E1, E2, or E3.

The effects of GnRHa with or without estrogens on BMD are shown graphically in Fig. 1. The administration of GnRHa significantly re-

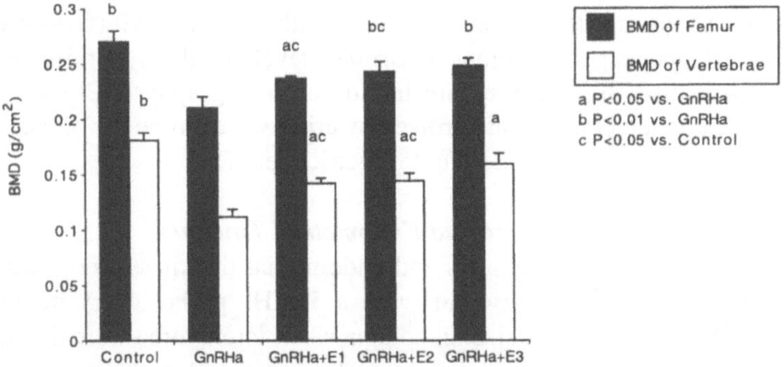

Fig. 1. The effects of GnRHa and estrogens on BMD of femur (*solid bars*) and lumbar vertebrae (*open bars*), measured by DXA. Values are mean±SEM (*a*: $p<0.05$ vs. GnRHa; *b*: $p<0.01$ vs. GnRHa; *c*: $p<0.05$ vs. control)

Table 1. Results of cancellous bone histomorphometry

Measurements	Control	GnRHa	GnRHa+E1	GnRHa+E2	GnRHa+E3
Bone volume (BV/TV, %)	17.0±1.3[b]	5.7±1.1	14.1±2.7[a]	12.6±1.9[a]	16.5±1.2[b]
Osteoid volume (OV/BV, %)	3.0±1.0[a]	5.5±1.0	0.7±0.2[b,c]	0.8±0.3[b,c]	0.9±0.2[b,c]
Trabecular thickness (Tb.Th, μm)	65.0±4.0	48.0±6.2	68.0±7.5[a]	67.5±9.5[a]	62.6±4.3[a]
Osteoid surface (OS/BS, %)	24.1±5.1	30.0±4.0	7.9±1.7[b,d]	9.9±2.8[b,c]	9.6±2.4[b,c]
Eroded surface (ES/BS, %)	42.2±4.1[b]	68.5±3.2	30.6±5.3[b]	33.3±5.2[b]	22.1±2.0[b,d]
Mineral apposition rate (MAR, μm/day)	0.6±0.1[b]	1.5±0.3	0.4±0.2[b]	0.5±0.2[b]	0.6±0.2[b]
Bone formation rate (BFR/BS, $\mu m^3/\mu m^2$/day)	0.2±0.03[b]	0.8±0.3	0.1±0.02[b]	0.1±0.03[b]	0.1±0.04[b]

Results are expressed as mean±SEM.
[a]$p<0.05$ vs. GnRHa.
[b]$p<0.01$ vs. GnRHa.
[c]$p<0.05$ vs. Control.
[d]$p<0.01$ vs. Control.

duced BMD of the femur and the lumbar vertebrae by 21.6% and 38.1%, respectively, compared with the control. BMD of rats given either E1 or E2 in combination with GnRHa was significantly higher than that of those given GnRHa alone. Rats administered E3 with GnRHa displayed BMD comparable with the control.

Data of cancellous bone histomorphometry are shown in Table 1. The bone volume of GnRHa-treated rats was diminished conspicuously, being 33.5% of the control. There was a slight decrease in trabecular thickness as a result of the GnRHa treatment. In contrast, osteoid volume, eroded surface, mineral apposition rate, and bone-formation rate in GnRHa-treated rats significantly increased by 83.3%, 62.3%, 114.3%, and 400.0%, respectively, compared with the control. There tended to be an increase in osteoid surface in GnRHa-treated rats although the difference between the GnRHa-treated group and the control was not significant. Concurrent treatment with E1, E2, or E3 attenuated these changes in bone parameters found in GnRHa-treated rats. A decrease in bone volume and trabecular thickness was eliminated by any one of the estrogen tested. It was noteworthy that osteoid volume and osteoid surface in estrogen-treated groups were much lower than those in the control. Eroded surface was reduced by estrogens to a lower level than that in GnRHa-treated group. Of the estrogens tested, E3 decreased eroded surfaces most remarkably, the value being by far lower than that of the control. The mineral-apposition rate and the bone-formation rate, which were stimulated by GnRHa treatment, were rather lower in the rats treated with E1, E2, or E3 than they were in the control.

8.3.2 Clinical Study

Baseline characteristics of the two groups were well-matched for age, height, weight, BMI, volume of the largest leiomyoma, BMD of the lumbar spine, and the values of serum E2 and biochemical markers of bone metabolism. The mean E2 concentrations significantly decreased to 10 pg/ml at 1 month of GnRHa therapy ($p<0.05$), and remained unaltered during the rest of the treatment period in both groups.

In the non-add-back group, lumbar spine (L2–L4) BMD was 1.196±0.061 g/cm^2 at the beginning of the treatment. BMD decreased to 96.5±3.7% and 92.5±1.7% ($p<0.05$) of the baseline at 4 and 6 months of treatment, respectively (Fig. 2), and then returned to levels similar to baseline (1.095±0.023 g/cm^2, 96.1±2.4% of baseline) 12 months after cessation of the treatment. In the add-back group, lumbar spine BMD was 1.211±0.079 g/cm^2 at pretreatment, and did not change significantly at 4 and 6 months of treatment (Fig. 2). There was a significant

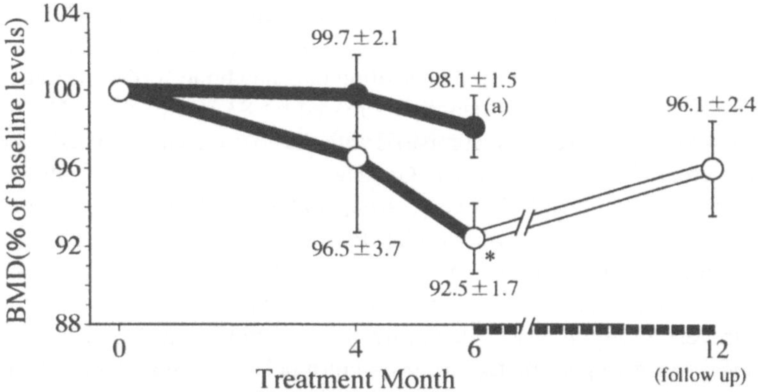

Fig. 2. Percent change in bone mineral density of lumbar spine (L2–L4) during the treatment period. *Closed circles*=add-back group; *Open circles*=non-add-back group; *Vertical lines* indicate the SEM; *$p<0.05$ vs. the baseline level; *(a)* $p<0.05$ vs. non-add-back group

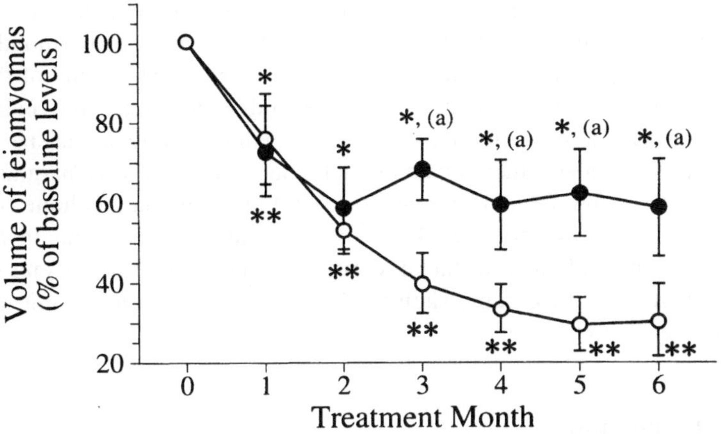

Fig. 3. Percent change in volume of uterine leiomyomas during the treatment period. *Closed circles*=add-back group; *Open circles*=non-add-back group; *Vertical lines* indicate the SEM; *$p<0.05$, **$p<0.01$ vs. the baseline level; *(a)* $p<0.05$ vs. non-add-back group.

difference ($p<0.05$) in BMD between the two groups at the end of the treatment.

As for the baseline levels of markers for bone metabolism in the non-add-back group, CrossLaps was 151.8±30.3 μg/mmol Cr, deoxypyridinoline 4.6±0.8 μmol/mmol Cr, osteocalcin 3.9±0.3 ng/ml, and B-ALP 50.5±4.9 IU/l. The levels of all the examined bone metabolic markers significantly increased at the end of treatment, compared with the baseline levels. The degree of an increase in CrossLaps levels (401.6±80.1% of baseline) was greater than in all other markers (deoxypyridinoline, 219.5±14.8%; B-ALP, 194.9±25.9%; osteocalcin, 211.2±9.3%). As for the baseline levels of markers for bone metabolism in the add-back group, CrossLaps was 160.1±40.5 μg/mmol Cr, deoxypyridinoline 5.2±1.3 μmol/mmol Cr, osteocalcin 3.9±0.6 ng/ml, and B-ALP 39.7±6.9 IU/l. An increase in the levels of all bone metabolic markers was blunted throughout the treatment and, therefore, their levels were significantly lower at the end of the treatment than it was for the non-add-back group.

At the beginning of treatment, the largest uterine leiomyoma volume was 374.1±228.2 cm^3 for the non add-back group and 244.4±81.5 cm^3

for the add-back group. In the non-add-back group, the mean leiomyoma volume was reduced to 53.6±5.9% of baseline at 2 months of GnRHa therapy, and further reduced to 31.3±9.3% at the end of the treatment (Fig. 3). In contrast, in the add-back group, the mean leiomyoma volume decreased to 59.1±10.3% of the pretreatment size at 2 months of GnRHa therapy; additional reduction in size was marginal. There was a significant difference ($p<0.05$) in leiomyoma volume between the two groups at 3, 4, 5, and 6 months of treatment. In all subjects, the endometrial thickness was less than 4 mm at the end of treatment. In both groups, vaginal bleeding did not occur during the treatment period.

8.4 Discussion

In the animal experiment, we demonstrated that E3 has unique tissue-selective effects in estrogen-deficient rats treated with GnRHa. E3 produced effects similar to E1 and E2 on BMD, bone metabolic markers, and cancellous bone turnover. Interestingly, E3 was shown to be less estrogenic than E1 and E2, as judged by uterine size. In the clinical study, we demonstrated that the significant decrease in the uterine leiomyoma volume that was achieved during the first 2 months of GnRHa alone could be maintained throughout the subsequent 4-month add-back therapy with E3 while preserving BMD of the lumbar spine.

It remains controversial whether E3 can effectively prevent the bone loss that affects postmenopausal women (Lindsay et al. 1979; Yang et al. 1995; Minaguchi et al. 1996; Itoi et al. 1997). Yang et al. (1995) reported that 2 mg/day of oral E3 administered for 2 years did not prevent bone loss in Chinese postmenopausal women. In another study in Scotland, postmenopausal osteoporosis was not prevented by administration of a high dose of E3, 12 mg/day, for 2 years (Lindsay et al. 1979). In contrast, a recent study conducted in Japan (Minaguchi et al. 1996) demonstrated that postmenopausal women given 2 mg/day E3 along with 800 mg/day calcium lactate showed a significant increase in lumbar spine BMD (1.79%) after 50 weeks of treatment. The bone-preserving effect was also observed in postmenopausal women treated for 24 months with 2 mg/day E3 plus 2.5 mg/day medroxyprogesterone acetate daily (Itoi et al. 1997). The discrepancy between these studies may be due to whether either calcium or progestin was coadministered or not. In both the human and animal experimentation, we have presented evidence supporting bone-protecting effects of E3 unsupplemented with other agents.

It is well known that bone loss caused by GnRHa treatment is associated with an increased bone turnover in women (Raven et al. 1994; Nakayama et al. 1997) and rats (Kurabayashi et al. 1993). In our study, there was a significant increase in serum or urinary levels of bone metabolic markers. These levels were significantly suppressed by the addition of E3. Thus, it is considered that the bone-protecting effect of E3 is achieved by the suppression of bone turnover.

As for cancellous bone histomorphometry in rats, GnRHa administration decreased bone volume and trabecular thickness, and increased bone resorption, as determined by eroded surface, and bone formation, judging from increases in osteoid volume, osteoid surface, mineral apposition rate, and bone formation rate. The addition of E1, E2, or E3 prevented all of the bone changes caused by GnRH administration. Notably, estrogens reduced the overall rate of bone turnover despite increasing bone volume and trabecular thickness. We reasoned that it is a subtle imbalance between bone formation and bone resorption, not the status of either bone formation or bone resorption, that determines bone volume.

Most of the effects of estrogen on reproductive tissues are believed to be mediated through estrogen receptors (ER). The ligand interaction with the receptor, the migratory step to the nucleus, and the interaction of the receptor–ligand complex with DNA have been described (Khovidhunkit and Shoback 1999; McKenna et al. 1999). There are two transcription-activating sites on the receptor, and activation of both appears to be required for full estrogen activity. In contrast to intensive investigations into the characterization of ER and the molecular mechanisms of estrogen actions in the reproductive tissues, the precise mechanism for the effects of estrogens on bone still has to be fully determined. Identification of ER on osteoblasts, osteoblast-like cells, and osteoclasts (Komm et al. 1988; Eriksen et al. 1988; Oursler et al. 1991; Arts et al. 1997) suggests that estrogens act directly on the bone through ER.

It is commonly believed that E3 is a safe estrogen, because it has little effect on endometrial tissue. It was documented that E3 given alone does not produce endometrial hyperplasia and other abnormalities, as judged by endometrial cytology and histology (Tzingounis et al. 1978; Grasso et al. 1982; Minaguchi et al. 1996). Our clinical data agreed with these findings in that E3 (4 mg/day) in conjunction with GnRHa did not stimulate endometrial proliferation as determined by ultrasonography. Furthermore, no breakthrough bleeding was noted in all the subjects on the add-back therapy, although uterine leiomyomas were partially affected by E3. In our animal experiment, E3 given with GnRHa successfully averted bone loss while its effect on uterine tissue was minimal compared with E1 and E2. In this sense, E3 can be regarded to be one of the selective estrogen receptor modulators (SERMs) (Cosman and Lindsay 1999). However, there have been conflicting reports on the effect of E3 on the endometrium (Tzingounis et al. 1978; Englund and Johansson 1980; Grasso et al. 1982; Punnonen and Söderström 1983; Montoneri et al. 1987; Minaguchi et al. 1996). It appears that the uterine growth-promoting effect of E3 depends on the route, dosage, and frequency of its administration (Clark et al. 1977; Heimer 1987).

E3 has biological activities in common with SERMs such as raloxifene, tamoxifen, and clomiphene, which are weak estrogen agonists on the uterus and potent agonists on the bone (Goulding and Fisher 1991; Goulding et al. 1992; Black et al. 1994; Evans et al. 1996; Sato et al. 1996; Jimenez et al. 1997; Khovidhunkit and Shoback 1999; Cosman and Lindsay 1999). However, E3 is effective in the management of climacteric symptoms (Tzingounis et al. 1980; Minaguchi et al. 1996), whereas the other SERMs have been reported to rather induce or deteriorate vasomoter symptoms such as hot flushes (Khovidhunkit and Shoback 1999; Cosman and Lindsay 1999). It appears that tissue-specific agonistic effects of these SERMs depend on the cell type, the ligand structure, ER subtypes, ER response element promoter context and transcriptional cofactors (Khovidhunkit and Shoback 1999; McKenna et al. 1999).

In conclusion, these results imply that, in light of its bone-protecting effects, associated with minimal uterotrophic activity, concurrent administration of E3 might be useful in long-term GnRHa treatment, thus offering a safer medical treatment strategy for women with estrogen-dependent disorders. The precise molecular mechanism of E3 actions in various estrogen target organs including the bone remains to be identified.

References

Adamson GD (1992) Treatment of uterine fibroids: current findings with gonadotropin-releasing hormone agonists. Am J Obstet Gynecol 166:746–751

Adashi EY (1994) Long-term gonadotropin-releasing hormone agonist therapy: the evolving issue of steroidal "add-back" paradigms. Human Reprod Update 9:1380–1397

Arts J, Kuiper GGJM, Janssen JMMF, Gustafsson JA, Löwik CWGM, Pols HAP, Van Leeuwen JPTM (1997) Differential expression of estrogen receptors α and β mRNA during differentiation of human osteoblast SV-HFO cells. Endocrinology 138:5067–5070

Black LJ, Sato M, Rowley ER, Magee DE, Bekele A, Williams DC, Cullinan GJ, Bendele R, Kauffman RF, Bensch WR, Frolik CA, Termine JD, Bryant HU (1994) Raloxifene (LY139481 HCI) prevents bone loss and reduces serum cholesterol without causing uterine hypertrophy in ovariectomized rats. J Clin Invest 93:63–69

Clark JH, Paszko Z, Peck EJ Jr (1977) Nuclear binding and retention of the receptor estrogen complex: relation to the agonistic and antagonistic properties of estriol. Endocrinology 100:91–96

Cosman F, Lindsay R (1999) Selective estrogen receptor modulators: clinical spectrum. Endocr Rev 20:418–434

Englund DE, Johansson EDB (1980) Endometrial effect of oral estriol treatment in postmenopausal women. Acta Obstet Gynecol Scand 59:449–451

Englund DE, Elamsson KB, Johansson EDB (1982) Bioavailability of oestriol. Acta Endocrinol 99:136–140

Eriksen EF, Colvard DS, Berg NJ, Graham ML, Mann KG, Spelsberg TC, Riggs BL (1988) Evidence of estrogen receptors in normal human osteoblast-like cells. Science 241:84–86

Evans GL, Bryant HU, Magee DE, Turner RT (1996) Raloxifene inhibits bone turnover and prevents further cancellous bone loss in adult ovariectomized rats with established osteopenia. Endocrinology 137:4139–4144

Goulding A, Fisher L (1991) Preventive effects of clomiphene citrate on estrogen-deficiency osteopenia elicited by LHRH agonist administration in the rat. J Bone Miner Res 6:1177–1181

Goulding A, Gold E, Feng W (1992) Tamoxifen in the rat prevents estrogen-deficiency bone loss elicited with the LHRH agonist, buserelin. Bone Miner 18:143–152

Grasso A. Baraghini F, Barbieri C, Vecchia ED, Previdi AM, Di Renzo GC, Volpe A (1982) Endocrinological features and endometrial morphology in climacteric women receiving hormone replacement therapy. Maturitas 4:19–26

Heimer GM (1987) Estriol in the postmenopause. Acta Obstet Gynecol Scand Suppl 139:1–23

Itoi H, Minakami H, Sato I (1997) Comparison of the long-term effects of oral estriol with the effects of conjugated estrogen, 1-α-hydroxyvitamin D_3 and calcium lactate on vertebral bone loss in early menopausal women. Maturitas 28:11–17

Jimenez MA, Magee DE, Bryant HU, Turner RT (1997) Clomiphene prevents cancellous bone loss from tibia of ovariectomized rats. Endocrinology 138:1794–1800

Khovidhunkit W, Shoback DM (1999) Clinical effects of raloxifene hydrochloride in women. Ann Intern Med 130:431–439

Komm BS, Terpening CM, Benz DJ, Graeme KA, Gallegos A, Korc M, Greene GL, O'Malley BW, Haussler MR (1988) Estrogen binding receptor mRNA, and biologic response in osteoblast-like osteosarcoma cells. Science 241:81–84

Kuiper GGJM, Carlsson B, Grandien K, Enmark E, Häggblad J, Nilsson S, Gustafsson JA (1997) Comparison of the ligand binding specificity and transcript tissue distribution of estrogen receptors α and β. Endocrinology 138:863–870

Kurabayashi T, Fujimaki T, Yasuda M, Yamamoto Y, Tanaka K (1993) Time-course of vertebral and femoral bone loss in rats administered gonadotrophin-releasing hormone agonist. J Endocrinol 138:115–125

Lindsay R, Hart DM, Maclean A, Garwood J, Clark AC, Kraszewski A (1979) Bone loss during oestriol therapy in postmenopausal women. Maturitas 1:279–285

McKenna NJ, Lanz RB, O'Malley BW (1999) Nuclear receptor coregulators: cellular and molecular biology. Endocr Rev 20:321–344

Melamed M, Castano E, Notides AC, Sasson S (1997) Molecular and kinetic basis for the mixed agonist/antagonist activity of estriol. Mol Endocrinol 11:1868–1878.

Minaguchi H, Uemura T, Shirasu K, Sato A, Tsukikawa S, Ibuki Y, Mizunuma H, Aso T, Koyama T, Nozawa S, Ohta H, Ikeda T, Kusuhara K, Ochiai K, Kato J, Kinoshita T, Tanaka K, Minagawa Y, Kurabayashi T, Fukunaga M (1996) Effect of estriol on bone loss in postmenopausal Japanese women: a multicenter prospective open study. J Obstet Gynaecol Res 22:259–265

Montoneri C, Zarbo G, Garofalo A, Giardinella S (1987) Effects of estriol administration on human postmenopausal endometrium. Clin Exp Obstet Gynecol 14:178–181

Nakayama H, Yano T, Sagara Y, Ando K, Kasai Y, Taketani Y (1997) Clinical usefulness of urinary CrossLaps as a sensitive marker of bone metabolism. Endocr J 44:479–484

Oursler MJ, Osdoby P, Pyfferoen J, Riggs BL, Spelsberg TC (1991) Avian osteoclasts as estrogen target cells. Proc Natl Acad Sci USA 88:6613–6617

Parfitt AM, Drezner MK, Glorieux FH, Kanis JA, Malluche H, Meunier PJ, Ott SM, Recker RR (1987) Bone histomorphometry: standardization of nomenclature, symbols, and units. J Bone Miner Res 2:595–610

Punnonen R, Söderström KO (1983) The effect of oral estriol succinate therapy on the endometrial morphology in postmenopausal women: the significance of fractionation of the dose. Europ J Obstet Gynecol Rprod Biol 14:217–224

Raven P, Bergqvist A, Hansen MA, Overgaard K, Christiansen C (1994) Treatment of endometriosis with the luteinizing hormone releasing hormone agonist nafarelin. Effect on bone turnover and bone mass. Menopause 1:11–17

Sato M, Rippy MK, Bryant HU (1996) Raloxifene, tamoxifen, nafoxidine, or estrogen effects on reproductive and nonreproductive tissues in ovariectomized rats. FASEB J 10:905–912

Schally AV (1989) The use of LHRH analogs in gynecology and tumor therapy. In: Belfort P, Pinotti JA, Eskes TKAB (eds) General gynecology, vol 6. Parthenon, Carnforth, pp 3–32

Schiff I, Tulchinsky D, Ryan KJ, Kadner S, Levitz M (1980) Plasma estriol and its conjugates following oral and vaginal administration of estriol to postmenopausal women: correlations with gonadotropin levels. Am J Obstet Gynecol 138:1137–1141

Surrey ES (1995) Steroidal and nonsteroidal "add-back" therapy extending safety and efficacy of gonadotropin-releasing hormone agonists in the gynecologic patient. Fertil Steril 64:673–685

Tzingounis VA, Aksu MF, Greenblatt RB (1978) Estriol in the management of the menopause. JAMA 239:1638–1641

Tzingounis VA, Aksu MF, Greenblatt RB (1980) The significance of oestriol in the management of the post-menopause. Acta Endorinol Suppl 233:45–50

Wheeler JM, Knittle JD, Miller JD (1992) Depot leuprolide versus danazol in treatment of women with symptomatic endometriosis. Am J Obstet Gynecol 167:1367–1371

Yang TS, Tsan SH, Chang SP, Ng HT (1995) Efficacy and safety of estriol replacement therapy for climacteric women. Clin Med J (Taipei) 55:386–391

9 Hypoestrogenic Bone Loss and HRT: The Predictive Value of Biochemical Markers of Bone Turnover

H. Minaguchi, M.G. Zhang, M. Taga

9.1 Introduction

In women, annual rates of bone loss are accelerated especially after menopause. It is well known that HRT following the menopause retards the rate of bone loss and substantially reduces the risk of fracture. Epidemiological evidence indicates that risk of hip fracture decreases with increasing duration of estrogen use. HRT taken for 10 years following menopause would reduce the risk of hip fracture by more than 50%, although undisciplined use for the prevention of osteoporosis will be costly. Bone loss in response to postmenopausal estrogen deficiency is reported to be heterogeneous, depending on each individual. Peak bone mass and subsequent bone loss are influenced by lifestyle as well as genetic factors. Furthermore, it has been argued that the effect of estrogens is reversed shortly after treatment is stopped, so that the most efficient method to prevent osteoporotic fracture is to identify women at risk and prevent osteoporosis.

Several markers for bone metabolism have been shown to reflect bone formation and resorption (Delmas 1991; Aloia et al. 1978; Lauffenburger et al. 1977; Nordin 1978). Some of these provide a semiquantitative index of bone resorption, but lack specificity and sensitivity (Marshall et al. 1996). The cross-links of mature collagen, pyridinoline (Pyr) and deoxypyridinoline (Dpyr), have been used to monitor bone resorption. They are formed nonenzymatically during maturation of extracellular collagen fibrils, they are then released by bone resorption, and excreted into the urine (Eyre 1992). Several studies have shown that Pyr and Dpyr are sensitive markers for bone resorption in the metabolic bone diseases characterized by increased bone turnover, including osteoporosis (Seibel et al. 1994) and malignancies affecting bone (Body and Delmas 1992; Coleman et al. 1992).

The type I collagen degradation products, urinary C-telopeptide (CTX) and N-telopeptide (NTX), are excreted as reproducible fractions of total bone-derived pyridinolines. New immunoassays for CTX and NTX provide a means to measure bone resorption (Rosen et al. 1994; Hanson et al. 1992; Garnero et al. 1994b; Gertz et al. 1994; Campodarve et al. 1995). Reported studies suggest that CTX and NTX are more sensitive and more specific indicators of bone resorption than pyridinolines measured by immunoassay are (Garnero et al. 1994a). NTX is reported to increase in peri- and postmenopausal women (Ebeling et al. 1996) and during ovarian suppression by GnRHa (Marshall et al. 1996; Dmowski et al. 1996). Conversely, it is reported to decrease as a result of antiresorptive therapy, such as hormone replacement (Rosen et al. 1997) or bisphosphonate administration (Rosen et al. 1994). Chesnut et al. (1997) reported that NTX provides a basis for predicting future bone loss in postmenopausal women and the probable efficacy of hormone replacement therapy. CTX is also reported to increase in postmenopausal women and to decrease in response to bisphosphonate treatment (Garnero et al. 1994b; Bonde et al. 1996). Garnero et al.

(1996) reported that CTX and Dpyr predict the susceptibility of elderly women to hip fractures. The study was undertaken to compare the clinical utility of CTX, NTX, and other markers for bone remodeling, in monitoring bone mass in a postmenopausal and gonadotropin-releasing hormone agonist (GnRHa) induced hypoestrogenic state and during the treatment of HRT.

9.2 Biochemical Markers

Urinary CTX was measured by ELISA (enzyme-linked immunosorbent assay) for CrossLaps (Osteometer A/S, Copenhagen, Denmark) according to the manufacturer's method. CrossLaps antibody was obtained by immunizing rabbits with the amino acid sequence specific for a part of the C-terminal telopeptide of the α_1 chain of type I collagen (Glu-Lys-Ala-His-Asp-Gly-Gly-Arg). The sensitivity was 50 μg/l. The intra- and interassay variabilities were % in the concentration range of the calibration curve. Duplicate measurements were performed for each urine sample and the values were corrected for creatinine (Cr), as measured by standard calorimetric technique.

Cross-linked N-telopeptide of type I collagen, NTX, was quantified directly in unextracted urine by ELISA with the use of a specific monoclonal antibody to NTX (Rosen et al. 1994). The ELISA kit was produced by Mochida Pharmaceutical Co. Ltd. (Tokyo, Japan) according to the modified method of Eyre (Rosen et al. 1994) and was supplied for clinical use. The values obtained are expressed as nanomolar bone-collagen equivalent (BCE) per millimolar Cr. The sensitivity of the assay was 20 nmol BCE/l. The intra- and interassay variabilities were 4.6% and 4.1%, respectively.

Serum osteocalcin (OC) was measured with a Mitsubishi Yuka Ltd (Tokyo, Japan) BGP IRMA kit with the use of a mouse monoclonal antibody to human OC. This procedure measures both the intact fragments and N- and C-terminal fragments. The sensitivity of the assay was 1.0 ng/ml. The intra- and interassay variabilities were 3.26% and 7.70%, respectively. Urinary Pyr and Dpyr were measured by HPLC with a modification of the method described by Uebelhart et al. (1990). An internal standard was used for the HPLC assay of pyridinolines. The values were expressed as nmol/mmol Cr. The sensitivity of both Pyr and

Dpyr was 4 pmol/ml. The intra- and interassay variabilities were 2.22% and 3.11% for Pyr, and 3.50% and 4.12% for Dpyr, respectively. Urine Hpr was measured according to standard colorimetric method. The intra- and interassay variabilities were 3.26% and 3.11%, respectively. Total alkaline phosphatase (Alp) was determined by standard enzymatic procedure. The intra- and interassay variabilities were 0.68% and 0.40%, respectively. Serum estradiol (E2) was measured by radioimmunoassay.

9.3 Statistical Analysis

Data were analyzed with a Stat View 2 (Abacus Concepts, Inc. Berkeley, Calif.) program on a Macintosh computer. Simple regression analysis was performed for each bone metabolic marker during the treatment. Stratification of sample data into quartiles was performed for descriptive purposes.

The statistical significance of correlation was determined with the F test. The statistical significance between the two groups was determined with one-way analysis of variance (ANOVA) followed by the Scheffe F test. The p values indicate the significance level of the difference between the means at each time point. Probability values less than 0.05 were considered statistically significant.

9.4 Changes in Urinary Excretion of CTX and NTX in Perimenopausal Women

A total of 92 women around the menopausal period were recruited for the study. The mean ages for pre- (n=19), peri- (n=22), and postmenopausal (n=51) women were 48.1±2.8, 48.6±2.1, and 52.9±3.9 (mean±SD), respectively.

Urinary concentration of Pyr, Dpyr, CTX, and NTX gradually increased during the course of pre-, peri-, and postmenopause. Urinary excretion of CTX (Fig. 1) and NTX (Fig. 2) in postmenopausal women was significantly higher than in premenopausal women. While Pyr and Dpyr also increased after menopause, the difference between premenopause and postmenopause was not statistically significant. The

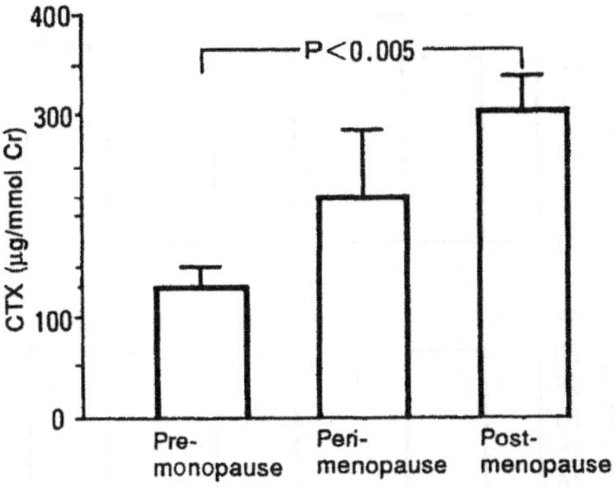

Fig. 1. Change (mean±SE) in urinary concentration of CTX during the menopausal transition. Perimenopausal women had experienced irregular menstrual frequency in the 12 months preceding entry to the study (Taga et al. 1998a, reproduced with permission from Horm Res 1998;49:86–90)

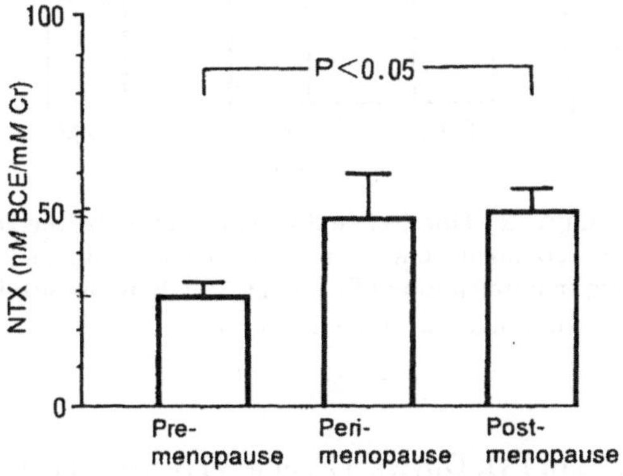

Fig. 2. Change (mean±SE) in urinary concentration of NTX during the menopausal transition. Perimenopausal women had experienced irregular menstrual frequency in the 12 months preceding entry to the study (Taga et al. 1998a, reproduced with permission from Horm Res 1998;49:86–90)

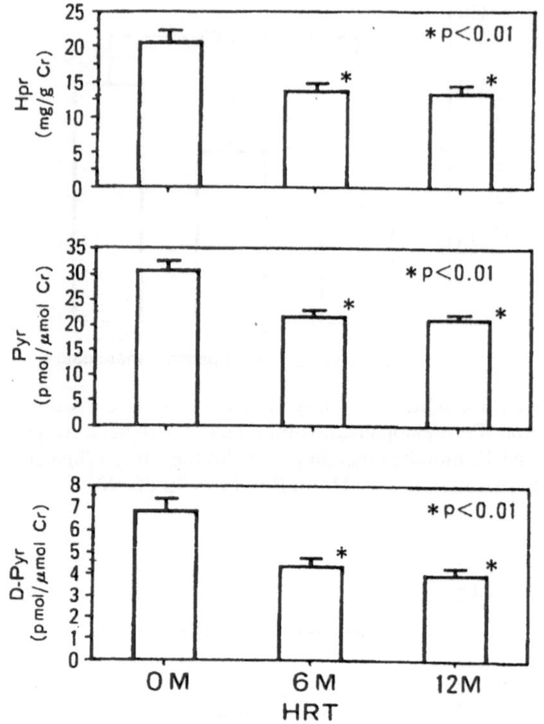

Fig. 3. Changes in urinary Hpr, Pyr, and Dpyr excretion before and at month 6 and month 12 of conjugate equine estrogen (0.625 mg) and medroxyprogesterone (2.5 mg) administration (Taga et al. 1998b, reproduced with permission from J Endocrinol Invest 21:154–159, 1998)

percent increases of Pyr, Dpyr, CTX, and NTX from the premenopausal levels to the perimenopausal ones were 29.6%, 58.5%, 68.1%, and 71.0%, respectively, and the levels were increased by 57.1%, 85.4%, 134%, and 75.1%, respectively, in postmenopause. The percent increases in Dpyr, CTX, and NTX, especially CTX, were greater than that of Pyr.

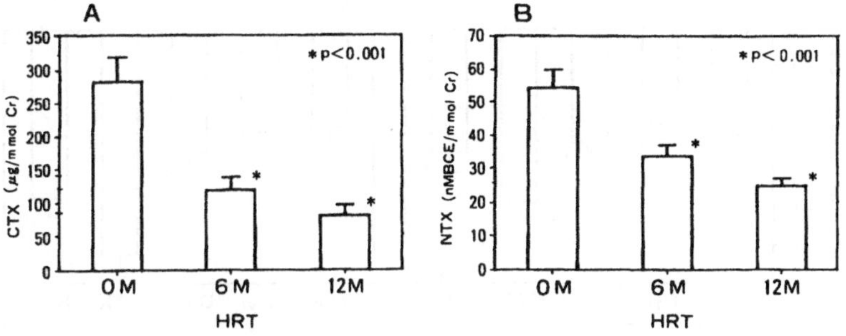

Fig. 4. Changes in urinary CTX (**A**) and NTX (**B**) excretions before and at month 6 and month 12 of conjugate equine estrogen (0.625 mg) and medroxyprogesterone (2.5 mg) administration (Taga et al. 1998b; reproduced with permission from J Endocrinol Invest 21:154–159, 1998)

9.5 Effect of Hormone Replacement Therapy in Postmenopausal Women on Urinary CTX and NTX

Thirty-three postmenopausal women, ages 43–66 years (52.9±5.3), with climacteric symptoms completed the study. Conjugated equine estrogen (Premarin 0.625 mg/day, Wyth) and medroxyprogesterone acetate (Provera 2.5 mg/day, Upjohn) were taken orally continuously for 12 months.

Figure 3 shows the changes in urinary Pyr, Dpyr, and Hpr during HRT. These values significantly decreased at 6 months after the initiation of HRT and remained lower than the baseline values during the 12 months of HRT. As shown in Fig. 4, both CTX(A) and NTX(B) also significantly decreased during HRT. While the values of Pyr, DPyr, and Hpr at 12 months of HRT did not differ from those at 6 months (Fig. 3), the levels of CTX and NTX had further decreased at 12 months (Fig. 4). In Fig. 5, the reduction rates of five bone resorption markers at 6 months (A) and 12 months (B) of HRT are compared.

As illustrated in Fig. 5, the mean percent decrease from the baseline, during HRT, of CTR was –48.3% (6 months) and –61.9% (12 months), and of NTX was –34.0% (6 months) and –46.5% (12 months). The reduction rates of CTX from the baseline were significantly greater than those of Pyr, Dpyr, and Hpr at 6 and 12 months of HRT, whereas that of NTX was significant at the 12th month but not at the 6th month (Fig. 5).

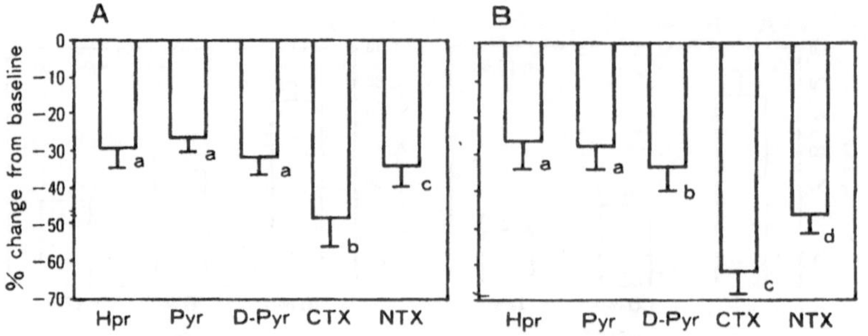

Fig. 5A,B. Comparison of the effect of HRT on the excretion of urine Hpr, Pyr, Dpyr, CTX, and NTX. Results represent the percent change from baseline in excretions of these bone resorption markers after 6 months (**A**) and 12 months (**B**) of conjugate equine estrogen (0.625 mg) and medroxyprogesterone (2.5 mg) administration. **A** *a* vs *b*, $p<0.01$; *b* vs *c*, $p<0.05$. **B** *a* vs *c*, *c* vs *b*, *c* vs *d*, $p<0.01$; *a* vs *d*, $p<0.05$. (Taga et al. 1998b; reproduced with permission from J Endocrinol Invest 21:154–159, 1998)

9.6 Relationship between Bone Resorption Markers and Bone Mineral Density during HRT

The pretreatment mean value of BMD in lumbar spine, determined by dual-energy X-ray absorptiometry (DXA) (XR-26, Norland), was 0.897±0.029 (mean±SD) g/cm^2. Figure 6 illustrates the relationship between the pretreatment value of CTX(A) or NTX(B) and baseline BMD. While the pretreatment values of Pyr, DPyr and Hpr did not correlate well with the baseline of BMD (data not shown), a negative correlation was evident in CTX (r=–0.338, $p<0.05$) and NTX (r=–0.298, p=–0.298, $p<0.05$).

9.7 Effects of GnRHa on BMD and Bone Metabolic Markers

Sixty-eight premenopausal, otherwise healthy Japanese women aged 22 to 45 years (34±5 years), with endometriosis or leiomyoma, were enrolled in this study. A therapeutic dose of long-acting GnRHa, either

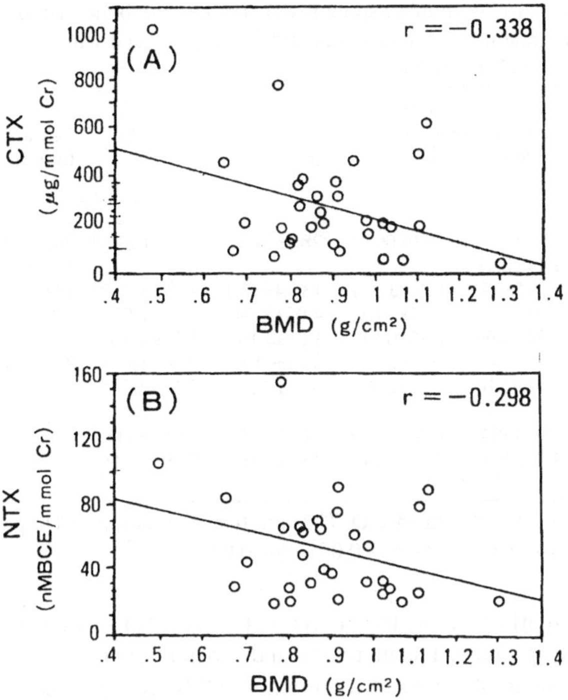

Fig. 6. Correlation between urinary CTX (**A**) or NTX (**B**) and lumbar spine (L^2–L^4) BMD at baseline (Taga et al. 1998b; reproduced with permission from J Endocrinol Invest 21:154–159, 1998)

1.8 mg of busereline microparticles (Hoechst, Germany) (n=50) or 3.6 mg of gosereline depot (Zeneca, UK) (n=18), was injected subcutaneously once a month for 24 weeks.

Changes in estradiol, lumbar spine BMD determined by DXA (Hologic, QDR 2000), and bone metabolic markers during 24 weeks of GnRHa treatment are presented in Table 1. The mean levels of serum estradiol were significantly suppressed by 4 weeks of treatment and remained below 30 pg/ml throughout the treatment. The mean value of BMD at 24 weeks of treatment was significantly lower than baseline (p<0.01) with a mean percentage bone loss of 3.79%. All the mean values for bone formation markers, Alp and OC, as well as for bone

Table 1. Changes in estradiol, lumbar spine BMD (L_2–L_4) and bone metabolic markers during 24 weeks of GnRHa treatment (Amama et al. 1998) (Reproduced with permission from J Clin Endocrinol Metab 83: 333–8, 1998)

	Stage of treatment: Pretreatment (*n*=68)	4 weeks (*n*=68)	12 weeks (*n*=68)	16 weeks (*n*=68)	24 weeks (*n*=68)	Percent change at 24 weeks from pretreatment
Estradiol (pg/ml)	61.8±122.0	28.9±60.2*	28.7±57.7*	16.8±62.7**	20.3±43.7**	–67.2±41.6
BMD (g/cm^2)	1.11±0.18				1.07±0.17**	–3.79±2.82
Alp (IU/l)	125.5±34.2	127.8±32.2	133.4±37.1	138.8±38.7*	153.7±36.3**	22.5±32.3
OC (ng/ml)	5.01±2.80	4.74±1.63	4.94±3.79	6.01±2.97*	7.14±2.80**	42.0±51.4
Hyp (nmol/g Cr)	1.34±0.99	1.36±0.91	1.82±1.81	1.21±1.16*	1.78±1.32*	32.8±73.4
Pyr (nmol/mmolCr)	33.5±14.4	39.1±23.1	33.4±25.6	37.4±15.7*	39.4±13.2**	17.6±37.7
Dpyr (nmol/mmol Cr)	5.62±2.14	5.93±2.39	6.13±4.54	7.63±3.38**	8.67±3.30**	54.3±48.5
CTX (μg/mmol Cr)	141.3±84.1	154.8±74.2	215.6±242.4	310.8±335.6*	370.6±223.4**	162.3±105.9
NTX (nmol BCE/mmol Cr)	41.0±19.0	42.9±18.1	45.2±19.0	50.9±23.1**	76.2±37.1**	85.8±76.1

The values are expressed as the mean±SD. The significance is compared to pretreatment values, with Student's *t*-test for paired data. *$p<0.05$. **$p<0.01$.

resorption markers, Hpr, Pyr, Dpyr, CTX, and NTX, increased as the treatment progressed. The increases in these markers from baseline were all significant at 24 weeks of treatment ($p<0.05$ for Hpr and $p<0.01$ for the other markers). Among these markers, the percent increase at 24 weeks of treatment from baseline was most marked in CTX (162%), followed by NTX (86%), Dpyr (54%), and OC (42%).

Table 2 shows the correlations between the percent decrease of BMD from baseline (bone loss) at 24 weeks of treatment and the seven biochemical markers (upper part of table) and the correlations among these markers (lower part of table) during GnRHa treatment. Strong negative correlations were observed between bone loss and CTX (r=–0.651) or NTX (r=–0.606). There were moderate or mild negative correlations between bone loss and the other markers except Hpr. Significant correlations were obtained among these seven markers themselves.

The patients were divided retrospectively into two groups according to the degree of bone loss after GnRHa treatment. For 36 patients, designated fast losers, bone loss was greater than the mean value of bone loss in all subjects, whereas for 32 patients, designated slow losers, bone

Table 2. The correlations between percent changes of lumbar spine BMD and biochemical markers (first line of table, "Percent change of BMD") and the correlations between markers (lower part of table) during GnRHa treatment (Amama et al. 1998) (Reproduced with permission from J Clin Endocrinol Metab 83: 333–8, 1998)

	Alp	OC	Hpr	Pyr	Dpyr	CTX	NTX
Percent change of BMD:	–0.394**	–0.328**	–0.160	–0.291**	–0.545**	–0.651**	–0.606**
Alp		0.309**	0.279**	0. 198**	0.367**	0.338**	0.391**
OC			0. 184**	0.223**	0.430**	0.401**	0.395**
Hpr				0. 173*	0.309**	0. 15 3*	0.368**
Pyr					0. 6 71	0.296**	0.379**
Dpyr						0.584**	0.666**
CTX							0.568**

*$p<0.05$. **$p<0.01$.

loss was less than the mean. Clinical and laboratory characteristics at baseline were not significantly different between the two groups. Figure 7 shows the comparisons of estradiol levels, percent changes of BMD, and percent changes of CTX and NTX from the pretreatment levels between the fast losers and slow losers during 24 weeks of treatment. Serum estradiol levels were similarly suppressed and remained under 30 pg/ml after 4 weeks of treatment in both groups; the two groups were not significantly different. The mean percent loss in BMD at 24 weeks of treatment was 5.87±0.26% in the fast losers, which was significantly greater ($p<0.0001$) than in the slow losers (1.45±0.34%). While CTX and NTX levels continued to increase during the treatment in both fast losers and slow losers, the percent increases at 24 weeks of treatment from baseline in the fast losers were significantly greater ($p<0.01$) than in the slow losers. In Fig. 8, the percent increases, from pretreatment levels, of the seven markers at 24 weeks of treatment is compared for the slow losers and the fast losers. The percent increases in Alp, Dpyr, CTX, and NTX in the fast losers were significantly higher than in the slow losers, whereas increases in OC, Hpr, and Pyr were not significantly different for the two groups.

To evaluate the discrimination power of the seven markers for bone loss in a hypoestrogenic state during GnRHa treatment, the z score for each of the seven markers at 24 weeks of treatment was calculated against the pretreatment level (Table 3). While the z scores were highest for CTX, NTX, and Dpyr, in that order, in both fast losers and slow

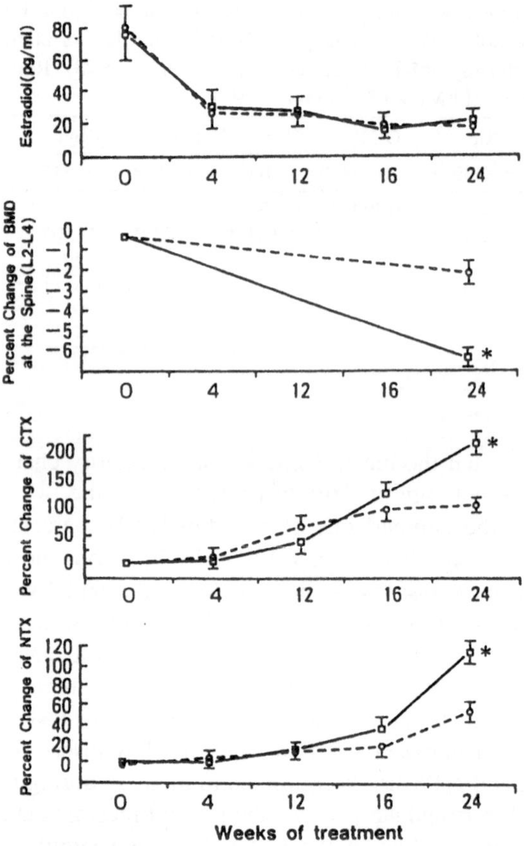

Fig. 7. Changes in serum level of estradiol and percent changes from pretreatment levels in lumbar spine BMD, CTX, and NTX in the fast losers and slow losers during 24 weeks of GnRHa treatment. *Solid lines* indicate the fast losers and *dotted lines* indicate the slow losers. The values are expressed as the mean±SE; *$p<0.01$, compared to the pretreatment values (Amama et al. 1998; reproduced with permission from J Clin Endocrinol Metab 83: 333–338, 1998)

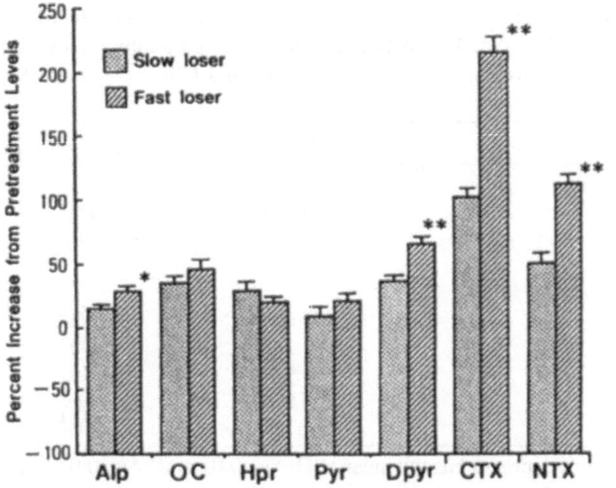

Fig. 8. Comparison, in percent increases from the pretreatment levels of seven markers, of the slow losers and fast losers of bone mass at 24 weeks of GnRH treatment. The values are expressed as the mean±SE; *$p<0.05$, **$p<0.01$, compared to the pretreatment values (Amama et al. 1998; reproduced with permission from J Clin Endocrinol Metab 83: 333–338, 1998)

losers, the highest z score ($p<0.05$) was observed for CTX (3.68±2.57) in the fast losers. NTX (2.53±1.93) and Dpyr (1.91±1.69) in the fast losers had moderate scores, whereas the z scores of Alp, OC, Hpr, and Pyr were low. The z scores of CTX, NTX, Dpyr, and Pyr in the fast losers were significantly greater than those in the slow losers, whereas the z scores of Alp, OC, and Hyr were not significantly different for the two groups.

To further analyze the relationship between bone loss due to GnRHa administration and each of CTX and NTX, we tried a stratification of CTX and NTX levels at 24 weeks of treatment into quartiles. As shown in Fig. 9A, the subjects in the highest quartile of CTX (475–655 µg/mmol Cr) experienced a significantly greater spinal BMD loss (2.1 times more) than the subjects within the lowest quartile of CTX (120 to approximately 226 µg/mmol Cr) (–5.72±0.50% vs –2.72±0.67%, $p<0.001$). Similarly, the subjects belonging to the highest (93 to ca. 145 nmol BCE/mmol Cr) and second highest quartile (66 to

Table 3. The *z* scores of markers in the slow losers and fast losers of bone mass at the 24th week of GnRHa treatment compared to pretreatment levels (Amama et al. 1998) (Reproduced with permission from J Clin Endocrinol Metab 82: 333–8, 1998)

	Slow losers	Fast losers
Bone formation		
Alp	0.68±[b,c]	1.02±[c]
OC	0.49±[c]	0.89±[c]
Bone resorption		
Hpr	0.60±[c]	0.37±[c]
Pyr	0.16±[c]	0.64±[c]*
Dpy r	0.92±[b,c]	1.91±[b]**
CTX	1.63±[a]	3.68±[a]**
NTX	1.14±[a,b]	2.53±[b]**

The values are expressed as the mean±SD; *$p<0.05$ and **$p<0.01$ as compared to the values of the slow losers. Means without a common superscript differ ($p<0.05$) between markers within each of the slow losers and fast losers to Duncan's multiple range test.

ca. 93 nmol BCE/mmol Cr) of NTX demonstrated 2.2 times ($p<0.001$) and 1.7 times ($p<0.05$) more spinal BMD loss, respectively, than the subjects within the lowest quartile of NTX (25 to ca. 50 nmol BCE/mmol Cr) (–5.87±0.46% and –4.44±0.73% vs –2.68±0.80%, respectively) (Fig. 9B). As illustrated in Fig. 9C, the subjects in the highest quartile of both CTX and NTX experienced 3.6 times more bone loss than the subjects in the lowest quartile of both CTX and NTX (–6.17±0.71% vs –1.69±0.88%, $p<0.001$). This magnitude of bone loss in the highest quartile of both CTX and NTX was also significantly greater ($p<0.01$) than in the highest quartile of either CTX or NTX, and 10.3% and 8.8% of all subjects were coincident in the highest and lowest quartile of both CTX and NTX, respectively.

Fig. 9. Percent bone loss at the lumbar spine by the quartiles of CTX (**A**), NTX (**B**), and by the highest and lowest quartiles of both CTX and NTX (**C**) at 24 weeks of GnRHa treatment. The values are expressed as the mean±SE; *$p<0.05$, **$p<0.001$, compared to Q1 (the lowest quartile) of either CTX, NTX, or both CTX and NTX (Amama et al. 1998; reproduced with permission from J Clin Endocrinol Metab 83: 333–8, 1998)

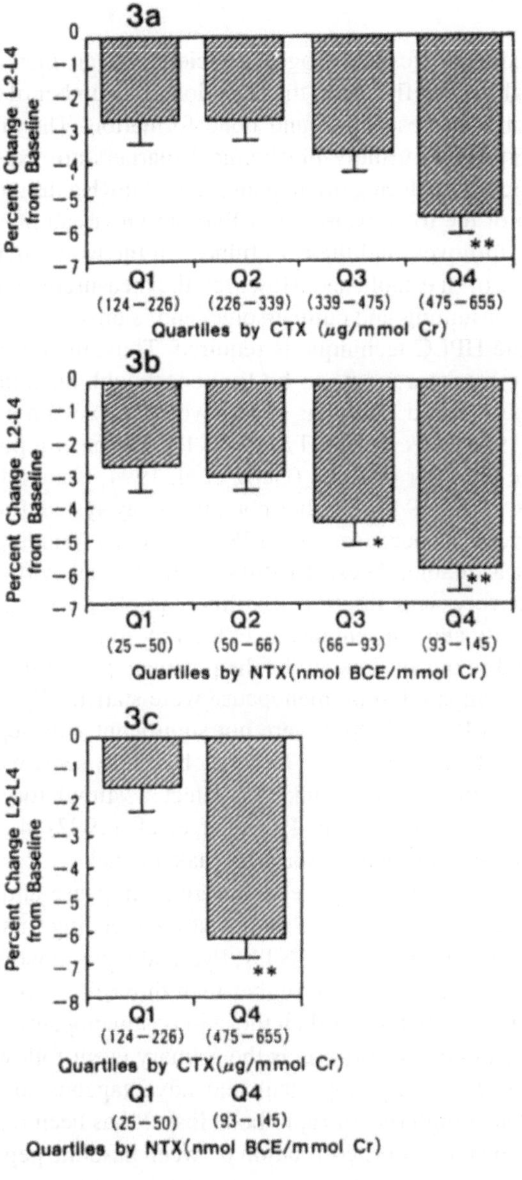
3a
Percent Change L2-L4 from Baseline
0
-1
-2
-3
-4
-5
-6
-7
**
Q1
(124-226)
Q2
(226-339)
Q3
(339-475)
Q4
(475-655)
Quartiles by CTX (μg/mmol Cr)
3b
Percent Change L2-L4 from Baseline
*
**
Q1
(25-50)
Q2
(50-66)
Q3
(66-93)
Q4
(93-145)
Quartiles by NTX(nmol BCE/mmol Cr)
3c
Percent Change L2-L4 from Baseline
-8
**
Q1
(124-226)
Q4
(475-655)
Quartiles by CTX(μg/mmol Cr)
Q1
(25-50)
Q4
(93-145)
Quartiles by NTX(nmol BCE/mmol Cr)

9.8 Discussion

The main characteristic of estrogen deficiency is an increased bone remodeling, which is reflected in the elevation of a number of biochemical markers of bone resorption and bone formation. Therefore, rapid changes in serum and urinary biochemical markers are expected in a hypoestrogenic state during menopause and GnRHa treatment. The determination of the total excretion of Pyr and Dpyr can be used as an index of bone turnover, and their usefulness in the assessment of bone metabolism has been established. However, the measurement of Pyr and Dpyr is time-consuming and cumbersome, and is unsuitable for routine use because the HPLC technique is required. Thus, the clinical use of these sensitive markers seems to be limited, though an immunoassay measuring free Pyr and Dpyr has been developed (Seydin et al. 1993; Risteli et al. 1993). Recently, CTX and NTX has been reported to be more sensitive than Pyr or Dpyr (Gertz et al. 1994; Campodarve et al. 1995; Garnero et al. 1994a). Furthermore, the assay system for CTX and NTX is convenient because an ELISA kit corresponding to each telopeptide is available. These ELISAs perform well with respect to precision, recovery, and dilution of urine samples, with greatly improved specificity and convenience over total pyridinoline analysis.

As indicated in our present study, the postmenopausal increments in CTX or NTX compared to premenopause were statistically significant, whereas those in Pyr and Dpyr were not significant; this suggests that the measurement of CTX or NTX could be more sensitive than the analysis of pyridinium cross-links to detect a stimulatory effect of menopause on bone resorption. Hassager et al. (1992) showed, in a longitudinal study, that bone resorption makers such as Pyr and Dpyr remained fairly constant in the years before menopause and started to increase about 6 months after the last menstrual bleeding. In the present study, urinary excretions of CTX, NTX, Pyr, and Dpyr in women whose menstruation was irregular were higher than those in women who had regular menstruation but lower than those in postmenopausal women.

An increase in bone resorption is the primary event following estrogen deficiency. It is very important and advantageous to accurately identify women at high risk of rapid bone loss. It has been reported that there is a significant inverse correlation between baseline peptide excretion and bone mineral density in the lumbar spine or the extent of bone loss (Gertz et al. 1994). Therefore, considering the superiority of CTX and NTX for detecting a subtle change in bone resorption in an early stage of ovarian follicular depletion, it is clinically important to utilize these telopeptides for the screening and management of women around the menopausal period.

Urinary CTX and NTX in postmenopausal women after 6 months of HRT significantly decreased to the level of premenopausal women. While the profiles of the changes in the excretion of these two telopeptides during HRT were almost consistent with those of Pyr, Dpyr, and Hpr, the reduction rates from baseline of CTX were significantly greater than those of Pyr, Dpyr, and Hpr at the 6th and 12th month of HRT. The magnitude of the percent decrease of NTX was also significantly greater at the 12th month of HRT. These results suggest that a suppressive effect of estrogen on bone resorption could be stronger in the excretion of C- or N-telopeptide than in the total excretion of Pyr. Thus, these two ELISAs provide a stronger index of bone resorption than is provided by Pyr, Dpyr, or Hpr.

As for the relationship between the bone resorption marker and bone mineral density, pretreatment values of CTX and NTX had a good negative correlation with the initial values of bone mineral density. Corresponding patterns were not found for Pyr, Dpyr, or Hpr. These results also indicate that these two telopeptides reflect more precisely the enhanced bone resorption state in postmenopausal women in whom bone mass begins to decline.

GnRHa, which temporarily down-regulates the pituitary–ovarian axis to induce a pseudomenopausal state, has recently been used to treat several gynecological diseases (Lemay et al. 1988; Faure and Lemay 1987; Maheux et al. 1985). GnRHa is effective in treating estrogen-dependent diseases such as endometriosis and leiomyoma (Meldrum et al. 1982). However, reduced bone density at some skeletal sites after treatment is a reported side effect caused by estrogen deficiency (Cann et al. 1987). Previous reports on biochemical markers during GnRHa treatment demonstrated an increase in urinary Hpr, Pyr, and Dpyr excretions (Gudmundsson et al. 1987; Steingold et al. 1987; van Leusden and Dogterom 1988), as well as an increase in urinary NTX excretion (Marshall et al. 1996; Dmowski et al. 1996). In the present study, all seven markers of both bone formation and bone resorption, including CTX and NTX, were significantly elevated during GnRHa therapy; this confirms that bone metabolic turnover is enhanced in the hypoestrogenic state induced by GnRHa. Among these markers, the percent increase from baseline at 24 weeks of treatment was most marked in CTX, followed by NTX. The calculated correlation coefficients between the percent change of BMD and the biochemical markers show that CTX and NTX correlate well with bone loss. Thus, CTX and NTX more sensitively detect increased bone resorption caused by a rapidly induced estrogen deficiency than pyridinoline cross-links do.

Bone loss in response to an estrogen deficiency is reported to be heterogeneous, depending on each individual (Slemenda et al. 1990; Hansen et al. 1991). In our study, a 14% variability was noted in bone loss during 24 weeks of GnRHa treatment, a result not unexpected, due to the various factors that affect bone loss. When the subjects were divided into two groups based on the degree of bone loss after GnRHa administration (fast losers and slow losers), the levels of spinal BMD as well as all markers before treatment were not different for the two groups. Estradiol levels were suppressed similarly in both groups during the treatment, and the extent of suppression was not significantly different for the two groups. However, at 24 weeks of treatment, the levels of Alp, Dpyr, CTX, and NTX in the fast losers were significantly higher than in the slow losers. Moreover, z scores of all bone resorption markers, except Hpr, in the fast losers were significantly higher than in the slow losers, while z scores of the bone formation markers, Alp and OC, did not differ for the two groups. Therefore, it is likely that bone resorption is more accelerated in the fast losers during treatment. While the z scores were highest for CTX, NTX, and Dpyr, in that order, at 24 weeks of treatment in both fast losers and slow losers, the highest z score among these markers was noted for CTX in the fast losers. The z score estimates the discrimination power of the assay, as it indicates a degree of difference under its baseline variability (Kushida et al. 1995). Thus, the new markers CTX and NTX, especially CTX, were considered to be better markers for monitoring the risk of bone loss during GnRHa therapy.

An accurate biochemical marker for the detection of postmenopausal women at risk for osteoporosis has been sought for years (Christiansen et al. 1987). NTX is reported to predict future bone loss and the therapeutic effects of hormone replacement therapy in postmenopausal women (Chesnut et al. 1997). In our current study, the subjects in the highest quartile of CTX, and in the highest and second highest quartiles of NTX at 24 weeks of treatment experienced 2.1, 2.2, and 1.7 times more bone loss, respectively, than those in the lowest quartiles. Furthermore, the subjects in the highest quartile of both CTX and NTX lost 3.4 times more bone mass than those in the lowest quartiles of both. Both CTX and NTX are cross-linked peptides derived from type I collagen; CTX originates in the carboxy-terminal nonhelical part (telopeptide) (Bonde et al. 1997) and NTX in the amino-terminal telopeptide (Rosen et al. 1994). The present study indicates that the concomitant measurements of both CTX and NTX may further increase the ability to evaluate bone loss. However, the present study indicates that neither CTX nor NTX can predict a risk of bone loss, because a significant difference between the bone resorption markers of the fast losers and the slow losers was observed only at the end of treatment. The failure to observe such a difference during treatment is possibly due to the short period of GnRHa administration and the abrupt onset of estrogen deficiency. The clinical applications of concomitant measurements of CTX and NTX in the management of postmenopausal women at risk of osteoporosis or osteopenia need to be studied further.

In conclusion, the present results indicate that CTX and NTX are useful and sensitive markers for bone resorption in a hypoestrogenic state. Both CTX and NTX can be used to monitor the changes in bone metabolism of postmenopausal women and during GnRHa treatment.

References

Aloia JF, Cohn SH, Zanzi I, Abesamis C, Ellis K (1978) Hydroxyproline peptides and bone mass in postmenopausal and osteoporotic women. J Clin Endocrinol Metab 47:314–318

Amama EA, Taga M, Minaguchi H (1998) The effect of gonadotropin-releasing hormone agonist on type I collagen C-telopeptide and N-telopeptide: the predictive value of biochemical markers of bone turnover. J Clin Endocrinol Metab 83:333-338

Body JJ Delmas PD (1992) Urinary pyridinium cross-links as markers of bone resorption in tumor-associated hypercalcemia. J Clin Endocrinol Metab 74:471–475

Bonde M, Fledelius C, Qvist P, Christiansen C (1996) Coated-tube radioimmunoassay for C-telopeptides of type I collagen to assess bone resorption. Clin Chem 42:1639–1644

Bonde M, Garnero P, Fledelius C, Qvist P, Delmas PD, Christiansen C (1997) Measurement of bone degradation products in serum using antibodies reactive with an isomerized form of an 8 amino acid sequence of the C-telopeptide of type I collagen. J Bone Miner Res 12:1028–1034

Campodarve I, Ulrich U, Bell N, et al (1995) Urinary N-telopeptide of type I collagen monitors bone resorption and may predict change in bone mass of the spine in response to hormone replacement therapy. J Bone Miner Res 10(Suppl):S182

Cann CE, Henzl M, Burry K, Andreyko J, Hanson F, Adamson GD, Trobough G, Henrichs L, Stewart G (1987) Reversible bone loss is produced by the GnRH agonist nafarelin. In: Cohn DV Martin TJ, and Meunier PJ (eds) Calcium regulation and bone metabolism: Basic and clinical aspects, vol 9. Elsevier Science, Amsterdam, pp 123–127

Chesnut CH 3rd, Bell NH, Clark GS, Drinkwater BL, English SC, Johnson CC Jr, Notelovitz M, Rosen C, Cain DF, Flessland KA, Mallinak NJ (1997) Hormone replacement therapy in postmenopausal women: urinary N-telopeptide of type I collagen monitors therapeutic effect and predicts response of bone mineral density. Am J Med 102:29–37

Christiansen C, Riis BJ, Rodbro P (1987) Prediction of rapid bone loss in postmenopausal women. Lancet May 16:1105–1107

Coleman RE, Houston S, James I, Rodger A, Rubens RD, Leonard RCF (1992) Preliminary results of the use of urinary excretion of pyridinium crosslinks for monitoring metastatic bone disease. Br J Cancer 65:766–768

Delmas PD (1991) Biochemical markers of bone turnover: methodology and clinical use in osteoporosis. Am J Med 91(Suppl 5B):59S-63S

Dmowski WP, Rana N, Pepping P, Cain DF, Clay TH (1996) Excretion of urinary N-telopeptides reflects changes in bone turnover during ovarian suppression and indicates individually variable estradiol threshold for bone loss. Fertil Steril 66:929–936

Ebeling PR, Atley LM, Guthrie JR, Burger HG, Dennerstein L, Hopper JL, Wark JD (1996) Bone turnover markers and bone density across the menopausal transition. J Clin Endocrinol Metab 81:3366–3371

Eyre D (1992) Editorial: New biomarkers of bone resorption. J Clin Endocrinol Metab 74:470A–470C

Faure N, Lemay A (1987) Ovarian suppression in polycystic ovarian disease during 6 months administration of a luteinizing hormone-releasing hormone (LHRH) agonist. Clin Endocrinol (Oxf) 27:703–713

Garnero P, Hausherr E, Chapuy MC, Marcelli C, Grandjean H, Muller C, Cormier C, Breart G, Meunier PJ, Delmas PD (1996) Markers of bone resorption predict hip fracture in elderly women: the EPIDOS prospective study. J Bone Miner Res 11:1531–1538

Garnero P, Gineyts E, Riou JP, Delmas PD (1994a) Assessment of bone resorption with a new marker of collagen degradation in patients with metabolic bone disease. J Clin Endocrinol Metab 79:780–785

Garnero P, Shih WJ, Gineyts E, Karpf DB, Delmas PD (1994b) Comparison of new biochemical markers of bone turnover in late postmenopausal osteoporotic women in response to alendronate treatment. J Clin Endocrinol Metab 79:1693–1700

Gertz BJ, Shao P, Hanson DA, Quan H, Harris ST, Genant HK, Chesnut CH 3rd, Eyre DR (1994) Monitoring bone resorption in early postmenopausal women by an immunoassay for cross-linked collagen peptides in urine. J Bone Miner Res 9:135–142

Gudmundsson JA, Ljunghall S, Bergquist C, Wide L, Nillius SJ (1987) Increased bone turnover during gonadotropin-releasing hormone superagonist-induced ovulation inhibition. J Clin Endocrinol Metab 65:159–163

Hansen MA, Overgaard K, Riis BJ, Christiansen C (1991) Role of peak bone mass and bone loss in postmenopausal osteoporosis: 12-year study. Brit Med J 303:961–964

Hanson DA, Weis MA, Bollen AM, Maslan SL, Singer FR, Eyre DR (1992) A specific immunoassay for monitoring human bone resorption: quantitation of type I collagen cross-linked N-telopeptides in urine. J Bone Miner Res 7:1251–1258

Hassager C, Colwell A, Assiri AMA, Eastell R, Russell RGG, Christiansen C (1992) Effect of menopause and hormone replacement therapy on urinary excretion of pyridinium cross-links: A longitudinal and cross-sectional study. Clin Endocrinol 37:45–50

Kushida K, Takahashi M, Kawana K, Inoue T (1995) Comparison of markers for bone formation and resorption in premenopausal and postmenopausal subjects, and osteoporosis patients. J Clin Endocrinol Metab 80:2447–2450

Lauffenburger T, Olah AJ, Dambacher MA, Guncaga J, Lentner C, Haas HG (1977) Bone remodeling and calcium metabolism: a correlated histomorphometric, calcium kinetic, and biochemical study in patients with osteoporosis and Paget's disease. Metabolism 26:589–606

Lemay A, Maheux R, Huot C, Blanchet J, Faure N (1988) Efficacy of intranasal or subcutaneous luteinizing hormone-releasing hormone agonist inhibition of ovarian function in the treatment of endometriosis. Am J Obstet Gynecol 158:233–236

Maheux R, Guilloteau C, Lemay A, Bastide A, Fazekas ATA (1985) Luteinizing hormone-releasing hormone agonist and uterine leiomyoma: a pilot study. Am J Obstet Gynecol 152:1034–1038

Marshall LA, Cain DF, Dmowski WP, Chesnut CH 3rd (1996) Urinary N-telopeptides to monitor bone resorption while on GnRH agonist therapy. Obstet Gynecol 87:350–354

Meldrum DR, Chang RJ, Lu J, Vale W, Rivier J, Judd HL (1982) "Medical oophorectomy" using a long-acting GnRH agonist – possible new approach to the treatment of endometriosis. J Clin Endocrinol Metab 54:1081–1083

Nordin BE (1978) Diagnostic procedures in disorders of calcium metabolism. Clin Endocrinol (Oxf) 8:55–67

Risteli J, Elonaa I, Niemi S, Novamo A, Risteli L (1993) Radioimmunoassay of type I collagen: A serum marker of bone collagen degradation. Clin Chem 39:635–640

Rosen HN, Dresner-Pollak R, Moses AC, Rosenblatt M, Zeind AJ, Clemens JD, Greenspan SL (1994) Specificity of urinary excretion of cross-linked N-telopeptides of type I collagen as a marker of bone turnover. Calcif Tissue Int 54:26–29

Rosen CJ, Chesnut CH 3rd, Mallinak NJ (1997) The predictive value of biochemical markers of bone turnover for bone mineral density in early postmenopausal women treated with hormone replacement or calcium supplementation. J Clin Endocrinol Metab 82: 1904–1910

Seibel MJ, Woitge H, Scheidt-Nave C, et al (1994) Urinary hydroxypyridinium crosslinks of collagen in population-based screening for overt vertebral osteoporosis: results of a pilot study. J Bone Miner Res 9:1433–1440

Seyedin EM, Kung VT, Daniloff YN, Helsey RP, Gomez B, Nielsen LA, Rosen HN, Zuk RF (1993) Immunoassay for urinary pyridinoline: the new marker for bone resorption. J Bone Miner Res 8:635–641

Slemenda CW, Hui SL, Longcope C, Wellman H, Johnston CC (1990) Predictors of bone mass in perimenopausal women. a prospective study of clinical data using photon absorptiometry. Ann Int Med 112:96–101

Steingold KA, Cedars M, Lu JKH, Randle D, Judd HL, Meldrum DR (1987) Treatment of endometriosis with a long-acting gonadotropin-releasing hormone agonist. Obstet Gynecol 69:403–411

Taga M, Shirashu K, Minaguchi H (1998a) Changes in urinary excretion of type-I collagen cross-linked C-telopeptide and N-telopeptide in perimenopausal women. Horm Res 49:86–90

Taga M, Uemura T, Minaguchi H (1998b) The effect of hormone replacement therapy in postmenopausal women on urinary C-telopeptide and N-telopeptide of type I collagen, new markers of bone resorption. J Endocrinol Invest 21:154–159

Uebelhart D, Gineyts E, Chapuy MC, Delmas PD (1990) Urinary excretion of pyridinium crosslinks: a new marker of bone resorption in metabolic bone disease. Bone Miner 8:87–96

van Leusden HA, Dogterom AA (1988) Rapid reduction of uterine leiomyoma with monthly injection of d-Trp6-GnRH. Gynecol Endocrinol 2:45–51

10 HRT and Climacteric Symptoms: Characteristics in Japanese Women

T. Aso

10.1 General Background to Menopause in Japanese Women

Menopause is the transition phase between the reproductive and nonreproductive stages. During this period, ovarian functions decline with aging, ovulation and menstrual cycles become irregular, and total endocrine function of the ovary diminishes and finally stabilizes. In Japanese women, this stage corresponds to 45 to 55 years of age. The menstrual status of healthy Japanese women between the ages of 40 to 60 is illustrated in Fig. 1 (Study Group on Reproduction and Endocrinology 1997). Approximately 20% of women in their early 40s had irregular menstrual cycles, more than half of the women became menopausal at the age of 52, and all had become menopausal at the age of 57. In general, these findings are not statistically different from those observed in women in other parts of the world.

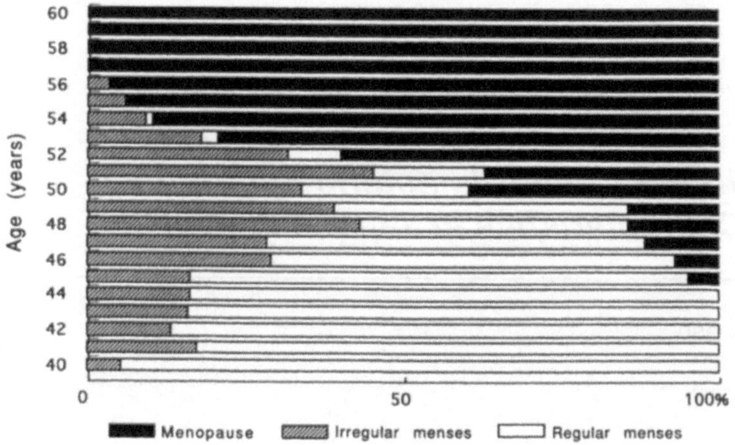

Fig. 1. The changes in the menstrual pattern of Japanese women between 40 and 60 years of age (from a study based on 3300 healthy women; Study Group on Reproduction and Endocrinology 1997)

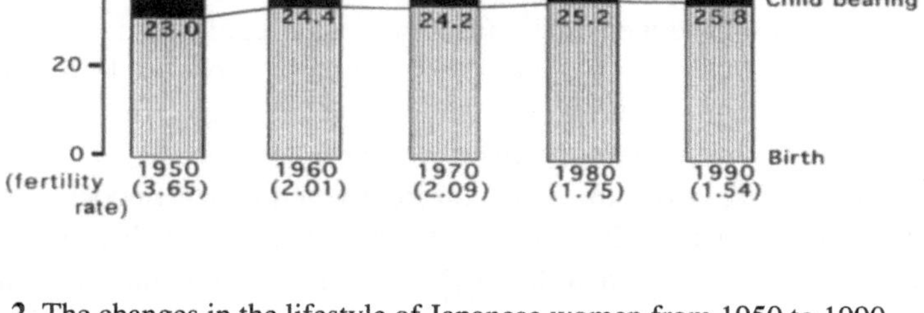

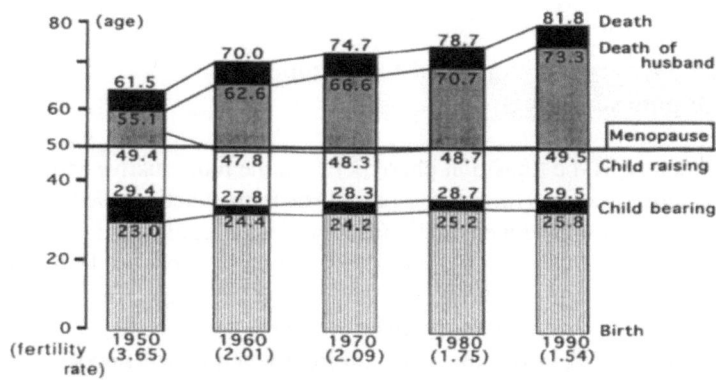

Fig. 2. The changes in the lifestyle of Japanese women from 1950 to 1990

Menopause is not only the end of the reproductive function, it is also a great turning point in a woman's life cycle. The changes in the lifestyle of women from 1950 to 1990 are depicted in Fig. 2 (Ministry of Health and Welfare, Population statistics 1992). The fertility rate in 1995 was 3.65 whereas that in 1990 was 1.54. In recent years, the childbearing age became higher and the number of children per family decreased significantly. On the other hand, the current life expectancies increased dramatically and women are expected to live 30 or more years after menopause. Based on these statistics, it is predicted that the proportion of the population above 65 years of age will increase up to 27% in 2025 in Japan (Ministry of Health and Welfare, Population census of Japan 1996). Thus, the prevention of nonphysiological aging processes by efficient medical interventions have high priority in the menopausal health care in Japan.

10.2 Climacteric Symptoms of Japanese Women

As indicated in cross-cultural studies, the incidence of climacteric symptoms taken as characteristic of menopause in Western countries cannot be assumed to be prevalent in Japan and other Asian countries (Boulet et al. 1994; Lock 1993; Payer 1991). The menopause of Japanese women is associated with fewer and less severe symptoms, and include psychological complaints such as anxiety, irritability, headache, depression, insomnia, and vasomotor symptoms. A previous study that compared the number of core symptoms of women in Japan, Canada, and the United States indicated that one-fourth of Japanese women had no core symptom and that approximately 10% had five and more symptoms. In contrast, 26% to 34% of the women in the other two countries had five and more symptoms (Lock 1997). A recent survey covering more than 3200 healthy Japanese women between the ages of 45 and 60 revealed that shoulder stiffness was the most common complaint (45%), and the percentage of women who were disturbed by moderate to severe vasomotor symptoms was approximately 25% (Fig. 3) (Study Group on Reproduction and Endocrinology 1997). A distinct different result was reported in a study on the menopausal women in Canada and the USA, in that hot flashes were observed in approximately 40% of them (Lock 1997). It seems that the major complaints of Japanese menopausal

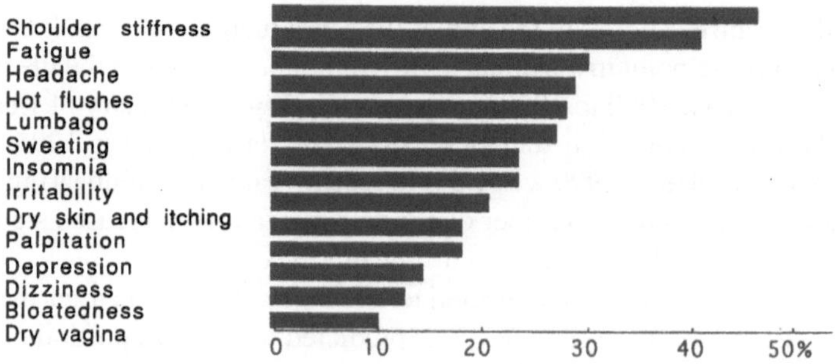

Fig. 3. The incidence of climacteric symptoms (to a moderate to severe degree) in Japanese perimenopausal women (Study Group on Reproduction and Endocrinology 1997)

women are not directly related to estrogen deficiency. Several factors have been postulated to influence the regional differences in climacteric symptoms. In particular, the lifestyle, including dietary habits, has been considered to be an important factor. The pattern of climacteric symptoms of Japanese women is possibly modified by such factors.

10.3 The Attitude of Japanese Women to HRT

The increasing amount of evidence has strongly confirmed the efficacy of HRT in the management of various problems in menopausal women. Since HRT was introduced for clinical use in Japan approximately 30 years ago, the effects of HRT have been appreciated mainly by gynecologists. Despite their efforts in providing information concerning the effectiveness of HRT, the number of women who accept this treatment is still limited. The general attitude of Japanese women to HRT is shown by the result of a survey, which was conducted among 2100 average, healthy Japanese women in perimenopause (Table 1) (Study Group on Reproduction and Endocrinology 1997). To the question on how they want to manage the climacteric symptoms, 40% to 50% of them answered that they preferred to manage the problems by improving their lifestyle. The proportion of the women who wanted to choose HRT was smaller than that of the women who wanted to have medical care not including HRT. It is not easy to indicate the reasons why Japanese

Table 1. The attitude of Japanese women towards managing menopause (*n*=2126) (Study Group on Reproduction and Endocrinology 1997)

	Pre-menopause	Menopause	Post-menopause
Improve on life style	51.5%	44.5%	41.3%
HRT	9.9%	17.2%	12.3%
Medical care except HRT	19.2%	23.8%	15.6%
Counseling only	15.5%	14.0%	16.2%

women do not accept HRT for the management of their climacteric symptoms. Alternative methods such as Kampo medicine have been widely adopted in Japan (Aso et al. 1996). The fear and hazards associated with HRT seem to make them hesitate to be treated with HRT, and a certain number of women are convinced that the administration of sex hormones after menopause it not natural and physiological, even if they want to have some treatment to improve their quality of life. It is also true that the attitude to HRT of the majority of physicians who are not specialized in menopausal care is the same as that of ordinary women. The genital bleeding induced by HRT is one of the common reasons for them to become confused and to discontinue the therapy.

10.4 HRT for Menopausal Health Care of Japanese Women in the Future

Because of their longevity, Japanese women are very concerned about menopausal health care. It is widely accepted that women in their reproductive years are protected from the risk of various diseases mainly because of the beneficial effects of estrogen. But once their estrogen levels decrease following menopause, the mortality from so-called adult diseases rises significantly. At the same time, the risks of osteoporosis and fracture, ischemic heart disease, dementia, and Alzheimer's disease, which are life-threatening and depress the quality of life tremendously, increase in the older women. Therefore, one should emphasize that strategies for prevention of these risks are urgently needed in Japan today. In addition to the short-term effects on climacteric symptoms, the long-term benefits of HRT play extremely important roles in the prevention of nonphysiological alterations in the aging process. To promote HRT in Japan, education in total health care for consumers and physicians is required. Furthermore, it should be established that the social health insurance system supports preventive medical care.

References

Aso T, Koyama T, Kaneko H, Seki M (1996) Alternative to hormone replacement therapy for menopausal management: possible role of Kampo medicine. In: Wren BG (ed) Progress in the management of the menopause. pp 71–77, Parthenon, London

Boulet MJ, Oddens BJ, Lehert P (1994) Climacteric and menopause in seven South-east Asian countries. Maturitas 19:157–176

Lock M (1993) Encounters with aging: mythologies of menopause in Japan and North America. University of California Press, Berkeley

Lock M (1997) Views of Japanese women on menopause: a discussion based on cultural differences with Canada and America. J Jpn Menopause Soc 5:53–59

Ministry of Health and Welfare (1992) Population statistics. Statistics and Information Department, Ministry of Health and Welfare, Japan

Ministry of Health and Welfare (1996) Population census of Japan. Statistics Bureau Management and Coordination Agency, Ministry of Health and Welfare, Japan

Payer L (1991) International health report. Menopause in various cultures. A portrait of the menopause. Parthenon, London

Study Group on Reproduction and Endocrinology (1997) The management of health care for the middle-aged women. Acta Obstet Gynecol Jpn 49:1019–1043

11 The Mechanisms of Bone Loss due to Estrogen Deficiency: Possible Role of Increased B-Lymphopoiesis in Bone Resorption

C. Miyaura

11.1 Role of Bone-Resorbing Cytokines in Bone Loss Induced by Estrogen Deficiency

An estrogen deficiency, caused by either menopause or removal of the ovaries, results in a marked bone loss by increased osteoclastic bone resorption. The pathologic bone loss in this condition can be restored by estrogen replacement therapy, but the mechanism underlying this phenomenon remains unknown. There appears to be a close relationship between bone remodeling and hemopoiesis in bone marrow (Suda et al. 1992). Osteoclast progenitors are thought to be of hemopoietic origin, and they are recruited from bone marrow (Suda et al. 1992). It is also suggested that several cytokines produced by bone marrow cells affect bone remodeling (Suda et al. 1992). Recent studies indicated the possible involvement of bone-resorbing cytokines such as interleukin (IL)-1, IL-6, and tumor necrosis factor α (TNFα) in bone loss due to estrogen deficiency.

In 1991, Pacifici et al. (1991) reported that an estrogen deficiency in oophorectomized women caused overproduction of IL-1 by peripheral blood monocytes, and that estrogen replacement therapy suppressed it. Kimble et al. (1994) reported that the administration of IL-1 receptor antagonist (IL-1ra), a specific competitor of IL-1, to ovariectomized (OVX) rats decreased both bone loss and bone resorption. Similar to IL-1, TNFα is also thought to be involved in the increased bone resorption due to estrogen deficiency. Administration of TNFα binding protein, a soluble type I TNF receptor which binds to TNFα and inhibits its biological functions, to OVX mice prevented bone loss by increased bone resorption (Kitazawa et al. 1994). Ammann et al. (1997) reported that OVX-induced changes in bone remodeling did not occur in transgenic mice that express soluble TNF receptor. These observations suggest that IL-1 and TNFα play a critical role in the pathogenesis of bone loss induced by estrogen deficiency.

Manolagas and his co-workers reported that IL-6 was involved in the stimulation of bone resorption induced by estrogen deficiency. In 1992, Girasole et al. (1992) reported that estrogen suppressed IL-6 production induced by IL-1 and/or TNFα in murine bone marrow stromal cells and osteoblastic cells. Jilka et al. (1992) reported that the increased osteoclastogenesis in OVX mice was prevented by giving them a neutralizing antibody against IL-6. Poli et al. (1994) have shown that OVX did not induce any change in bone mass in the IL-6-knockout mice. These results suggest that IL-6 also has an important role in the stimulation of osteoclastogenesis in estrogen deficiency.

To explore the endogenous bone-resorbing factors involved in estrogen deficiency, we examined the bone-resorbing activity present in the supernatant fraction of mouse bone marrow collected from OVX mice (Miyaura et al. 1995). Bone marrow was flushed out from tibiae and femora, and bone marrow supernatants were prepared by centrifugation. Bone-resorbing activity of bone marrow supernatants was determined by the release of calcium from fetal mouse long bones. The endogenous bone-resorbing activity was much greater in OVX mice than in sham mice. Antibodies against IL-1α neutralized the bone-resorbing activity completely, but the concentration of IL-1α present in bone marrow supernatants was not enough to induce bone resorption. Antibodies against IL-6 also neutralized the bone-resorbing activity, but only partially. The bone-resorbing activity of bone marrow supernatants was significantly decreased by adding indomethacin; this suggests that prostaglandins are involved in the mechanism of bone resorption induced by estrogen deficiency. Using these bone marrow supernatants, Kawaguchi et al. (1995) reported that the marrow supernatants collected from OVX mice greatly stimulated cyclooxygenase (COX)-2 mRNA expression and PGE synthesis in mouse calvarial cultures. Recently, we have shown that simultaneous treatment with submaximal doses of IL-1α and IL-6 with soluble IL-6 receptor caused marked induction of osteoclast formation and COX-2-dependent PGE synthesis (Tai et al. 1997). These results suggest that several bone-resorbing cytokines, such as IL-1, TNFα, IL-6, and PGs act cooperatively in inducing bone resorption during estrogen deficiency.

11.2 Estrogen Deficiency Stimulates B-Lymphopoiesis in Mouse Bone Marrow

There is a close relationship between bone remodeling and hemopoiesis in bone marrow. A microenvironment supported by bone marrow stromal cells plays an important role not only in hemopoietic cells but also in bone remodeling. While investigating the role of estrogen in bone remodeling, we found that an estrogen deficiency induced by OVX caused a marked increase in bone marrow cells. To examine the effect of estrogen on hemopoiesis, we characterized the increased population of bone marrow cells after OVX. In OVX mice, the number of B220-positive B lymphocytes was selectively increased 2–4 weeks after surgery (Masuzawa et al. 1994). The total number of myeloid cells and granulocytes did not change appreciably. When OVX mice were treated with estrogen, the increased B lymphopoiesis returned to normal. B220-positive cells were classified into two subpopulations, $B220^{low}$ and $B220^{high}$. The majority of the $B220^{low}$ cells were negative for the μ chain of Ig M, whereas most of the $B220^{high}$ cells were μ-chain positive.

OVX selectively increased the precursors of B lymphocytes identified by the B220^{low} Ig M μ-chain-negative phenotype, suggesting that an estrogen deficiency results in a marked accumulation of pre-B cells in bone marrow (Masuzawa et al. 1994). In contrast, B-lymphopoiesis is significantly reduced in bone marrow of pregnant mice with high plasma levels of estrogen (Medina et al. 1993).

We also examined the effects of estrogen on the growth and differentiation of B cells in vitro. When bone marrow cells were co-cultured with bone-marrow-derived stromal cells (ST2), precursors of B lymphocytes were selectively grown and differentiated into mature B cells on the ST2 cell layers. Treatment with 17β-estradiol strikingly suppressed the generation of B220-positive B lymphocytes in this culture system (Masuzawa et al. 1994). Smithson et al. (1995) also reported that estrogen suppressed stromal-cell-dependent B lymphopoiesis in an in vitro culture system. These results are consistent with the in vivo data; this indicates selective accumulation of pre-B cells in bone marrow during estrogen deficiency. The results also suggest that estrogen plays an important role in the regulation of B lymphocyte development in mouse bone marrow.

Like estrogen deficiency, androgen deficiency also results in bone loss by stimulating osteoclastic bone resorption. Orchidectomy (ORX) stimulated B-lymphopoiesis and caused a marked accumulation of pre-B cells in the bone marrow of male mice (Chaki et al. 1996; Wilson et al. 1995). The characteristics of the accumulated B cells in ORX mice were quite similar to those in OVX mice. Both the increased B-lymphopoiesis and the decreased bone mass were completely restored by the treatment with 17β-estradiol (Chaki et al. 1996). It is known that testosterone converts to estrogen by local aromatase. These observations suggest that testosterone is metabolized to estrogen by local aromatase, and that estrogen regulates B-lymphopoiesis in male mice as well.

11.3 Relationship between B-Lymphopoiesis and Bone Metabolism

IL-7 is a growth factor responsible for early B cell differentiation, and selectively supports the growth of pre-B cells in vitro and in vivo (Namen et al. 1988; Lee et al. 1989; Morrissey et al. 1991). To examine

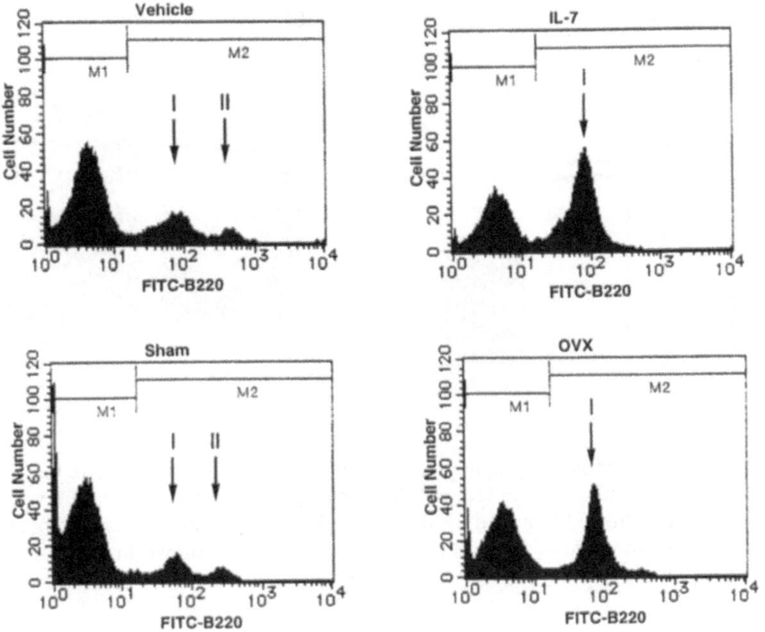

Fig. 1. Comparison of the increased B-lymphopoiesis in bone marrow of IL-7-treated mice and OVX mice. Results of the flow cytometric analysis of B220-positive bone marrow cells of the vehicle- and IL-7-treated mice on day14, and the sham-operated and OVX mice 2 weeks after operation were compared. Note that the proportion of B220-positive cells was markedly increased in the IL-7-treated mice and in the OVX mice

the possible correlation between stimulated B-lymphopoiesis and bone loss, female mice were treated with IL-7. When mice were treated with IL-7, the number of B220-positive B cells was selectively and markedly increased on days 4 to 20. In the vehicle-treated normal mice, most of the B220-positive B cells were B220^{high} Ig M μ-chain positive. In the IL-7-treated mice, the B220^{high} Ig M μ-chain-positive population was reduced, whereas the B220^{low} Ig M μ-chain-negative population was markedly increased. Therefore, it is concluded that treatment with IL-7 and estrogen deficiency both induce B-lymphopoiesis, resulting in an accumulation of pre-B cells in mouse bone marrow (Fig. 1) (Miyaura et al. 1997). The uterine weight was markedly reduced in OVX mice,

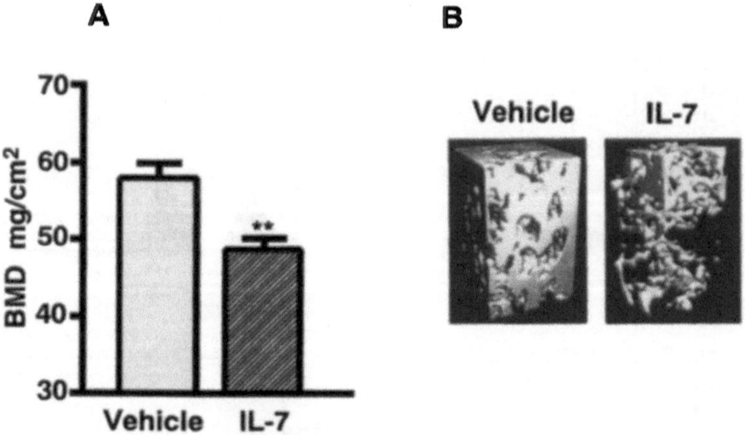

Fig. 2A,B. Increased B-lymphopoiesis by IL-7 treatment induces bone loss. **A** BMD was measured in the femoral distal metaphysis collected from the vehicle- and IL-7-treated female mice on day 20. **B** The three-dimensional trabecular bone architecture of femoral metaphysis of the mice shown in A was analyzed by μCT scanning

whereas IL-7 treatment had no effect on it, indicating that the IL-7-induced B-lymphopoiesis is not related to ovarian function.

To examine the relationship between the perturbation of B-lymphopoiesis and the change in bone mass, the density of mineralized cancellous bone was analyzed in the femora of IL-7-treated mice. Treatment with IL-7 significantly reduced the bone mineral density (BMD) in the femoral metaphysis, compared with that in the vehicle-treated mice (Miyaura et al. 1997). The degree of bone loss induced by IL-7 was quantitatively similar to that induced by estrogen deficiency (Fig. 2). Histological analysis of femoral sections showed not only a marked loss of cancellous bone, but also an increase in the osteoclast number in the secondary spongiosa of the distal metaphysis in IL-7-treated mice, compared with the vehicle-treated controls. In addition, morphometric analysis with micro-computed-tomography (μCT) and its three-dimensional analysis revealed that, as in OVX mice, bone resorption was stimulated in the IL-7-treated mice (Fig. 2). IL-7 did not stimulate bone resorption in fetal mouse long bones in organ cultures

(Onoe et al. 1996). Therefore, the change in bone marrow microenvironment accompanying the increased B-lymphopoiesis may stimulate osteoclastic bone resorption, to result in bone loss in vivo.

11.4 Effects of Selective Estrogen Receptor Modulator (SERM) on B-Lymphopoiesis and Bone Metabolism

Recently, attention has been focused on the tissue-specific estrogen agonists. They are estrogen-related compounds that selectively act on bone without exhibiting substantial estrogenic action in the uterus. Some of the synthetic compounds such as raloxifene preferentially act on bone and the cardiovascular systems as an agonist, whereas they antagonize the effects of estrogen in reproductive tissue (Black et al. 1994). These compounds are known as selective estrogen receptor modulators (SERMs) and are currently available for the prevention of osteoporosis. Although the protective effect of raloxifene in bone loss is established, the mode of action of raloxifene as a SERM is controversial. Raloxifene also has potent estrogenic activity for bone, but has minimal estrogenic activity for the uterus.

To examine the effects of raloxifene on B-lymphopoiesis and bone resorption, OVX mice were subcutaneously, with a mini-osmotic pump, given either estrogen or raloxifene for 2–4 weeks. Reduced uterine weight in OVX mice was completely restored by 17β-estradiol (E2). Some 300-fold higher doses of raloxifene increased the uterine weight of OVX mice, but only slightly (Fig. 3). After OVX, the number of B220-positive pre-B cells was markedly increased in the bone marrow. The increased B-lymphopoiesis was prevented not only by E2 but also by raloxifene (Fig. 3) (Onoe et al. in press). In OVX mice, the trabecular bone volume of the femoral distal metaphysis was markedly reduced, and both E2 and raloxifene similarly restored it. Like estrogen deficiency, androgen deficiency induced by orchidectomy(ORX) also resulted in a marked bone loss and increased B-lymphopoiesis. Both E2 and raloxifene prevented the changes in ORX mice (Onoe et al. in press). These results indicate that raloxifene exhibits estrogenic actions in bone and bone marrow to prevent bone loss and regulate B-lymphopoiesis without inducing estrogenic action in the uterus. In addition, Ishimi et al. (1999) reported that genistein, a soybean isoflavone, exhib-

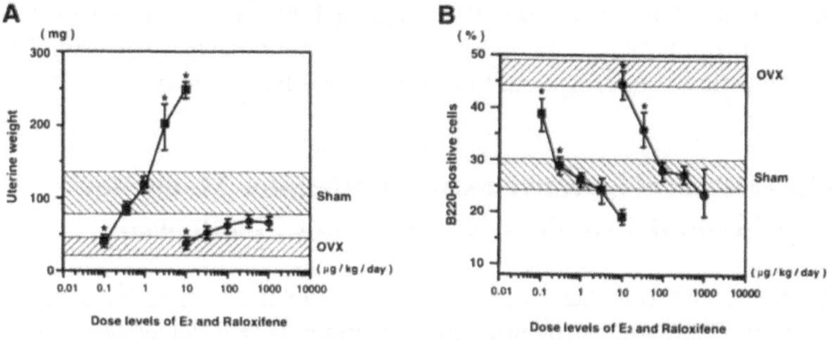

Fig. 3A,B. Effects of raloxifene and 17β-estradiol on the uterine weight and the population of B220-positive cells in bone marrow in OVX mice. Mice were sham-operated or OVX, and some of the OVX mice were treated with various doses of raloxifene (*filled circles*) or 17β-estradiol (*filled squares*). Two weeks later, the uterine weight was measured (**A**) and bone marrow cells were collected to examine the expression of B220-positive cells analyzed by flow cytometry (**B**)

its estrogenic action in bone and in bone marrow to regulate B-lymphopoiesis and prevent bone loss without exhibiting estrogenic action in uterus. It has been reported that genistein selectively bind to ERβ rather than ERα. These studies with SERMs indicate the close relationship between B-lymphopoiesis and bone metabolism.

11.5 Possible Involvement of ODF/RANKL/OPGL in Bone Loss Induced by Estrogen Deficiency

Recently, osteoclast differentiation factor (ODF) has been cloned as a crucial factor for osteoclast formation, and was expressed on the surface of osteoblasts and bone marrow stromal cells treated with bone-resorbing factors such as IL-1, 1α,25-dihydroxyvitamin D_3 [1α,25$(OH)_2D_3$], parathyroid hormone (PTH), and PGE2 (Suda et al. 1999; Yasuda et al. 1998). ODF was identical to RANKL (receptor activator of NF-kB ligand), TRANCE (TNF-related activation-induced cytokine) and OPGL (osteoprotegerin ligand), which were cloned independently (Anderson et al. 1997; Wong et al. 1997; Lacey et al. 1998). ODF is essential for the differentiation of osteoclast progenitors into mature

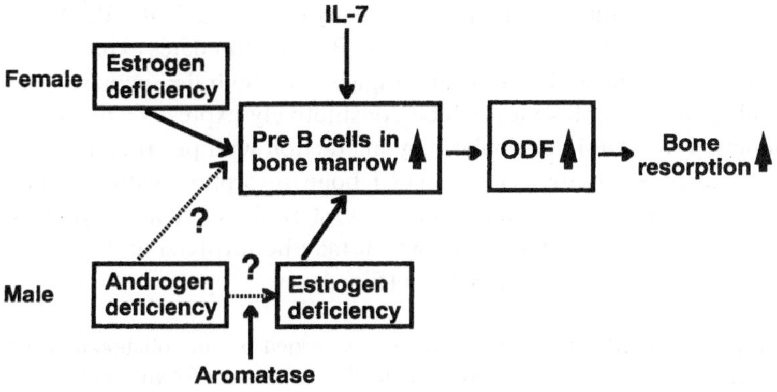

Fig. 4. A possible relation between increased B-lymphopoiesis and stimulated bone resorption by sex steroid deficiency. Increased B-lymphopoiesis induced not only by sex steroid deficiency but also by IL-7 administration commonly stimulates osteoclastic bone resorption, resulting in a marked bone loss in female and male mice. This suggests that the increased B-lymphopoiesis may be involved in the mechanism of stimulated bone resorption in both OVX and ORX mice. Stimulated expression of ODF was suggested in the process of stimulated bone resorption

osteoclasts (Suda et al. 1999). To study the possible involvement of ODF in the increased bone resorption due to estrogen deficiency, we examined expression of ODFmRNA in trabecular bone and bone marrow B-lymphocytes in OVX mice.

At 2–4 weeks after OVX, trabecular bone and bone marrow cells were collected from tibia for RT-PCR to examine the expression of ODF. OVX markedly induced expression of ODFmRNA in trabecular bone and bone marrow, compared with sham mice (Asami et al. 1999). When B-cells were isolated from bone marrow, with the use of B220-coated magnetic micro-beads, more than 98% of the isolated cells were B220-positive B-lymphocytes, and the isolated B-cells constitutively expressed ODFmRNA(Asami et al. 1999). When the isolated B-cells were co-cultured with mouse primary osteoblasts, the expression of ODF was greatly enhanced. To compare the level of expression of ODFmRNA in B-cells and osteoblasts, B-cells were separated from osteoblasts cell layer after the co-culture. The level of ODFmRNA in B-cells did not change before and after the co-culture. In contrast, osteoblasts co-cultured with B-cells greatly expressed ODFmRNA, suggesting that cell-to-cell interaction with B-cells stimulates expression of ODF in osteoblasts. These results suggest that both the increase in the number of B-lymphocytes, which constitutively express ODF, and the induction of ODF in osteoblasts by interaction with pre-B cells appear to be responsible for OVX-induced bone resorption with enhanced osteoclastogenesis. In conclusion, sex steroid deficiency stimulates bone marrow B-lymphopoiesis, which may be involved in the mechanism of stimulated bone resorption (Fig. 4).

Acknowledgments. The present study was carried out in collaboration with Drs. T. Suda (Showa University), M. Inada, T. Asami (Tokyo University of Pharmacy and Life Science), K. Ikuta, K. Maki (Kyoto University), M. Ito (Nagasaki University), T. Masuzawa, Y. Onoe, H. Ohta (Keio University), and O. Chaki (Yokohama City University). The author also thanks Drs. T. Sato (Daiichi Pharmaceutical Co. Ltd.) and Y. Nagai (Kureha Chemical Industry.) for their helpful assistance in histological studies.

References

Ammann P, Rizzoli R, Bonjour J, Bourrin S, Meyer JM, Vassalli P, Garcia I (1997) Transgenic mice expressing soluble tumor necrosis factor-receptor are protected against bone loss caused by estrogen deficiency. J Clin Invest 99:1699–1703

Anderson DM, Maraskovsky E, Billingsley WL, Dougall WC, Tometsko ME, Roux ER, Teepe MC, DuBose RF, Cosman D, Galibert L (1997) A homologue of the TNF receptor and its ligand enhance T-cell growth and dendritic-cell function. Nature 390:175–179

Asami T, Inada M, Mizunuma H, Ibuki Y, Miyaura C (1999) Expression of osteoclast differentiation factor (ODF/RANKL/OPGL) in trabecular bone and bone marrow B-lymphocytes in OVX mice. J Bone Miner Res 14:S152

Black LJ, Sato M, Rowley ER, Magee DE, Bekele A, Williams DC, Cullinian GJ, Bendele R, Kauffman RF, Bensch WR, Frolik CA, Termine JD, Bryant HU (1994) Raloxifene (LY139481 HCl) prevents bone loss and reduces serum cholesterol without uterine hypertrophy in ovariectomized rats. J Clin Invest 93:63–69

Chaki O, Miyaura C, Seo H, Gorai I, Minaguchi H, Suda T (1996) Androgen deficiency stimulates B-lymphopoiesis in bone marrow and induces bone loss in male mice: Comparison of the effects of estrogen and androgen. In:

Papapoulos SE, Lips P, Pols HAP, Johnston CC, Delmas PD (eds) Osteoporosis 1996. Proceeding of the 1996 World Congress on Osteoporosis, Elsevier, Amsterdam, pp 37–41

Girasole G, Jilka RL, Passeri G, Boswell S, Boder G, Williams DC, Manolagas SC (1992) 17β-Estradiol inhibits interleukin-6 production by bone marrow-derived stromal cells and osteoblasts in vitro: A potential mechanism for the antiosteoporotic effect of estrogens. J Clin Invest 9:883–891

Ishimi Y, Miyaura C, Ohmura M, Onoe Y, Sato T, Uchiyama Y, Ito M, Wang X, Suda T, Ikegami S (1999) Selective effects of genistein, a soybean isoflavone, on B-lymphopoiesis and bone loss caused by estrogen deficiency. Endocrinology 140:1893–1900

Jilka RL, Hangoc G, Girasole G, Passeri G, Williams DC, Abrams JS, Boyce B, Broxmeyer H, Manolagas SC (1992) Increased osteoclast development after estrogen loss: Mediation by interleukin-6. Science 257:88–91

Kawaguchi H, Pilbeam CC, Vargas SJ, Morse EE, Lorenzo JA, Raisz LG (1995) Ovariectomy enhances and estrogen replacement inhibits the activity of bone marrow factors that stimulate prostaglandin production in cultured mouse calvariae. J Clin Invest 96:539–548

Kimble RB, Vannice JL, Bloedow DC, Thompson RC, Hopfer W, Kung VT, Brownfield C, Pacifici R (1994) Interleukin-1 receptor antagonist decreases bone loss and bone resorption in ovariectomized rats. J Clin Invest 93:1959–1967

Kitazawa R, Kimble RB, Vannice JL, Kung VT, Pacifici R (1994) Interleukin-1 receptor antagonist and tumor necrosis factor binding protein decrease osteoclast formation and bone resorption in ovariectomized mice. J Clin Invest 94:2397–2406

Lacey DL, Timms E, Tan HL, Kelley MJ, Dunstan CR, Burgess T, Elliott R, Colombero A, Elliott G, Scully S, Hsu H, Sullivan J, Hawkins N, Davy E, Capparelli C, Eli A, Qian YX, Kaufman S, Sarosi I, Shalhoub V, Senaldi G, Guo J, Delaney J, Boyle WJ (1998) Osteoprotegerin ligand is a cytokine that regulates osteoclast differentiation and activation. Cell 93:165–176

Lee G, Namen AE, Gillis S, Ellingsworth LR, Kincade PW (1989) Normal B cell precursors responsive to recombinant murine IL-7 and inhibition of IL-7 activity by transforming growth factor-β. J Immunol 142:3875–3883

Masuzawa T, Miyaura C, Onoe Y, Kusano K, Ohta H, Nozawa S, Suda T (1994) Estrogen deficiency stimulates B lymphopoiesis in mouse bone marrow. J Clin Invest 94:1090–1097

Medina KL, Smithson G, Kincade PW (1993) Suppression of B lymphopoiesis during normal pregnancy. J Exp Med 178:1507–1515

Miyaura C, Kusano K, Masuzawa T, Chaki O, Onoe Y, Aoyagi M, Sasaki T, Tamura T, Koishihara Y, Ohsugi Y, Suda T (1995) Endogenous bone-resorbing factors in estrogen deficiency: Cooperative effects of IL-1 and IL-6. J Bone Miner Res 10:1365–1373

Miyaura C, Onoe Y, Inada M, Maki K, Ikuta K, Ito M, Suda T (1997) Increased B-lymphopoiesis by interleukin 7 induces bone loss in mice with intact ovarian function: Similarity to estrogen deficiency. Proc Natl Acad Sci USA 94:9360–9365

Morrissey PJ, Conlon P, Charrier K, Braddy S, Alpert A, Williams D, Namen AE, Mochizuki D (1991) Administration of IL-7 to normal mice stimulates B-lymphopoiesis and peripheral lymphadenopathy. J Immunol 147:561-568

Namen AE, Lupton S, Hjerrild K, Wignall J, Mochizuki DY, Schmierer A, Mosley B, March CJ, Urdal D, Gillis S, Cosman D, Goodwin RG (1988) Stimulation of B cell progenitors by cloned murine IL-7. Nature 333:571–573

Onoe Y, Miyaura C, Kaminakayashiki T, Nagai Y, Noguchi K, Chen QR, Seo H, Ohta H, Nozawa S, Kudo I, Suda T (1996) IL-13 and IL-4 inhibit bone resorption by suppressing cyclooxygenase-2-dependent prostaglandin synthesis in osteoblasts. J Immunol 156:758–764

Onoe Y, Miyaura C, Ito M, Ohta H, Nozawa S, Suda T (in press) Comparative effects of estrogen and raloxifene on B-lymphopoiesis and bone loss induced by sex steroid deficiency in mice. J Bone Miner Res

Pacifici R, Brown C, Puscheck E, Friedrich E, Slatopolsky E, Maggio D, McCracken R, Avioli LV (1991) Effect of surgical menopause and estrogen replacement on cytokine release from human blood mononuclear cells. Proc Natl Acad Sci USA 88:5134–5138

Poli V, Balena R, Fattori E, Markatos A, Yamamoto M, Tanaka H, Ciliberto G, Rodan GA, Costantini F (1994) Interleukin-6 deficient mice are protected from bone loss caused by estrogen depletion. EMBO J 13:1189–1196

Smithson G, Medina K, Ponting I, Kincade PW (1995) Estrogen suppresses stromal cell-dependent lymphopoiesis in culture. J Immunol 155:3409–3417

Suda T, Takahashi N, Martin TJ (1992) Modulation of osteoclast differentiation. Endocr Rev 13:66–80

Suda T, Takahashi N, Udagawa N, Jimi E, Gillespie MT, Martin TJ (1999) Modulation of osteoclast differentiation and function by the new members of the tumor necrosis factor receptor and ligand families. Endocrine Rev 20:345–357

Tai H, Miyaura C, Pilbeam CC, Tamura T, Ohsugi Y, Koishihara Y, Kubodera N, Kawaguchi H, Raisz LG, Suda T (1997) Transcriptional induction of cyclooxygenase-2 in osteoblasts is involved in interleukin-6-induced osteoclast formation. Endocrinology 138:2372–2379

Wilson CA, Mrose SA, Thomas DW (1995) Enhanced production of B lymphopoiesis after castration. Blood 85:1535–1539

Wong BR, Rho J, Arron J, Robinson E, Orlinick J, Chao M, Kalachikov S, Cayani E, Bartlett FS III, Frankel WN, Lee SY, Choi Y (1997) TRANCE is a novel ligand of the tumor necrosis factor receptor family that activates c-Jun N-terminal kinase in T cells. J Biol Chem 272:25190–25194

Yasuda H, Shima N, Nakagawa N, Yamaguchi K, Kinosaki M, Mochizuki S, Tomoyasu A, Yano K, Goto M, Murakami A, Tsuda E, Morinaga T, Higashio K, Udagawa N, Takahashi N, Suda T (1998) Osteoclast differentiation factor is a ligand for osteoprotegerin/osteoclastogenesis-inhibitory factor and is identical to TRANCE/RANKL. Proc Natl Acad Sci USA 95:3597–3602

Subject Index

Ernst Schering Research Foundation Workshop

Editors: Günter Stock
Monika Lessl

Vol. 1 *(1991)*: Bioscience ⇋ Society – Workshop Report
Editors: D. J. Roy, B. E. Wynne, R. W. Old

Vol. 2 (1991): Round Table Discussion on Bioscience ⇋ Society
Editor: J. J. Cherfas

Vol. 3 (1991): Excitatory Amino Acids and Second Messenger Systems
Editors: V. I. Teichberg, L. Turski

Vol. 4 (1992): Spermatogenesis – Fertilization – Contraception
Editors: E. Nieschlag, U.-F. Habenicht

Vol. 5 (1992): Sex Steroids and the Cardiovascular System
Editors: P. Ramwell, G. Rubanyi, E. Schillinger

Vol. 6 (1993): Transgenic Animals as Model Systems for Human Diseases
Editors: E. F. Wagner, F. Theuring

Vol. 7 (1993): Basic Mechanisms Controlling Term and Preterm Birth
Editors: K. Chwalisz, R. E. Garfield

Vol. 8 (1994): Health Care 2010
Editors: C. Bezold, K. Knabner

Vol. 9 (1994): Sex Steroids and Bone
Editors: R. Ziegler, J. Pfeilschifter, M. Bräutigam

Vol. 10 (1994): Nongenotoxic Carcinogenesis
Editors: A. Cockburn, L. Smith

Vol. 11 (1994): Cell Culture in Pharmaceutical Research
Editors: N. E. Fusenig, H. Graf

Vol. 12 (1994): Interactions Between Adjuvants, Agrochemical and Target Organisms
Editors: P. J. Holloway, R. T. Rees, D. Stock

Vol. 13 (1994): Assessment of the Use of Single Cytochrome P450 Enzymes in Drug Research
Editors: M. R. Waterman, M. Hildebrand

Vol. 14 (1995): Apoptosis in Hormone-Dependent Cancers
Editors: M. Tenniswood, H. Michna

Vol. 15 (1995): Computer Aided Drug Design in Industrial Research
Editors: E. C. Herrmann, R. Franke

Vol. 16 (1995): Organ-Selective Actions of Steroid Hormones
Editors: D. T. Baird, G. Schütz, R. Krattenmacher

Vol. 17 (1996): Alzheimer's Disease
Editors: J.D. Turner, K. Beyreuther, F. Theuring

Vol. 18 (1997): The Endometrium as a Target for Contraception
Editors: H.M. Beier, M.J.K. Harper, K. Chwalisz

Vol. 19 (1997): EGF Receptor in Tumor Growth and Progression
Editors: R. B. Lichtner, R. N. Harkins

Vol. 20 (1997): Cellular Therapy
Editors: H. Wekerle, H. Graf, J.D. Turner

Vol. 21 (1997): Nitric Oxide, Cytochromes P 450, and Sexual Steroid Hormones
Editors: J.R. Lancaster, J.F. Parkinson

Vol. 22 (1997): Impact of Molecular Biology and New Technical Developments in Diagnostic Imaging
Editors: W. Semmler, M. Schwaiger

Vol. 23 (1998): Excitatory Amino Acids
Editors: P.H. Seeburg, I. Bresink, L. Turski

Vol. 24 (1998): Molecular Basis of Sex Hormone Receptor Function
Editors: H. Gronemeyer, U. Fuhrmann, K. Parczyk

Vol. 25 (1998): Novel Approaches to Treatment of Osteoporosis
Editors: R.G.G. Russell, T.M. Skerry, U. Kollenkirchen

Vol. 26 (1998): Recent Trends in Molecular Recognition
Editors: F. Diederich, H. Künzer

Vol. 27 (1998): Gene Therapy
Editors: R.E. Sobol, K.J. Scanlon, E. Nestaas

Vol. 28 (1999): Therapeutic Angiogenesis
Editors: J.A. Dormandy, W.P. Dole, G.M. Rubanyi

Vol. 29 (2000): Of Fish, Fly, Worm and Man
Editors: C. Nüsslein-Volhard, J. Krätzschmar

Vol. 31 (2000): Advances in Eicosanoid Reseach
Editors: C.N. Serhan, H.D. Perez

Supplement 1 (1994): Molecular and Cellular Endocrinology of the Testis
Editors: G. Verhoeven, U.-F. Habenicht

Supplement 2 (1997): Signal Transduction in Testicular Cells
Editors: V. Hansson, F. O. Levy, K. Taskén

Supplement 3 (1998): Testicular Function:
From Gene Expression to Genetic Manipulation
Editors: M. Stefanini et al.

Supplement 4 (2000): Hormone Replacement Therapy and Osteoporosis
Editors: J. Kato, H. Minaguchi, Y. Nishino

Supplement 5 (1999): Interferon:
The Dawn of Recombinant Protein Drugs
Editors: J. Lindenmann, W.D. Schleuning